Synthese Library

Studies in Epistemology, Logic, Methodology, and Philosophy of Science

Volume 537

The aim of *Synthese Library* is to provide a forum for the best current work in the methodology and philosophy of science and in epistemology, all broadly understood. A wide variety of different approaches have traditionally been represented in the Library, and every effort is made to maintain this variety, not for its own sake, but because we believe that there are many fruitful and illuminating approaches to the philosophy of science and related disciplines.

Special attention is paid to methodological studies which illustrate the interplay of empirical and philosophical viewpoints and to contributions to the formal (logical, set-theoretical, mathematical, information-theoretical, decision-theoretical, etc.) methodology of empirical sciences. Likewise, the applications of logical methods to epistemology as well as philosophically and methodologically relevant studies in logic are strongly encouraged. The emphasis on logic will be tempered by interest in the psychological, historical, and sociological aspects of science. In addition to monographs *Synthese Library* publishes thematically unified anthologies and edited volumes with a well-defined topical focus inside the aim and scope of the book series. The contributions in the volumes are expected to be focused and structurally organized in accordance with the central theme(s), and should be tied together by an extensive editorial introduction or set of introductions if the volume is divided into parts. An extensive bibliography and index are mandatory.

Alexander Ehmann

Worlds in a Stochastic Universe: On the Emergence of World Histories in Tychistic Bohmian Mechanics

Springer

Chapter 1
Introduction and Overview

This book is largely based on my doctoral research at Lingnan University, Hong Kong, and my research as an Ernst Mach fellow at the University of Salzburg. It develops a detailed account of the emergence of for all practical purposes (FAPP) continuous, quasi-classical world histories, containing FAPP-continuous macroscopic phenomena, from the discontinuous, stochastic micro dynamics of Tychistic Bohmian Mechanics (TBM). TBM is a non-relativistic quantum theory, recently put forward by Patrick Dürr and me.[1] It results from excising the guiding equation from standard Bohmian Mechanics (BM) and reinterpreting the quantum equilibrium hypothesis as a stochastic guidance law for the random actualization of configurations of Bohmian particles. On TBM, there are no continuous trajectories linking up individual configurations. Instead, individual configurations are actualized independently of each other, carving out the decoherence-induced branching structure of the universal wave function (WF). Yet, by contrast to the Everett interpretation, branches of the universal wave function are not actualized in parallel, i.e., all at the same time. Rather, world branches, on TBM, are actualized sequentially.

Along the way of developing my approach of emergent world histories, I will be discussing several, interrelated topics. First, I will give a short introduction to TBM (Chap. 2), using standard BM as the background against which it stands out. I will discuss BM's solution of the eminent measurement problem (Sect. 2.1) and its empirical equivalence with orthodox QM (Sect. 2.2). The transition from standard BM to TBM will be made by excising the guiding equation from the former and reinterpreting the quantum equilibrium hypothesis as a stochastic law, governing TBM's fundamentally random, discontinuous particle dynamics (Sect. 2.3). I will argue that TBM is empirically equivalent with standard BM, and thus with QM,

[1] See Dürr and Ehmann (2021). For various reasons, I cannot fully reproduce this paper here. Instead, I will outline TBM in a significantly shortened and simplified manner appropriate for our present purposes. For a more detailed account, the reader is referred to the paper mentioned.

A. Ehmann, *Worlds in a Stochastic Universe: On the Emergence of World Histories in Tychistic Bohmian Mechanics*, Synthese Library 537,
https://doi.org/10.1007/978-3-032-14632-8_1

despite its lack of continuous, deterministic particle trajectories. Since many proponents of BM take the latter to be a primitive ontology (PO) theory, and regard this as somewhat crucial, I will argue that TBM can be made compatible with the PO approach, given certain restrictions on the interpretation of its WF (Sect. 2.4). TBM to some extent resembles Bell's "Everett (?) theory". The latter was immediately rejected by its creator on the charge of leading to the problem of temporal solipsism (Sect. 2.5). I will locate the root of this problem in Bell's identification of instantaneous configurations with worlds: on his construal, each world contains one, and only one, configuration. Accordingly, each world is likely to be instantiated only once. Barbour's time-capsule approach, at first sight, suggests itself as a possible solution to the problem of temporal solipsism. Time capsules are a special type of configurations: they contain consistent records of past events that, according to Barbour, suffice for the reconstruction of consistent histories from within individual, static configurations. However, Barbour's time-capsule approach provides no account of how temporally remote configurations are linked. Like Bell, Barbour identifies individual configurations—albeit special ones—with worlds. Thus, his approach isn't suitable to solve the problem of temporal solipsism within the TBM framework (Sect. 2.6).

In order to tackle this issue, I will propose to adopt a notion of worlds different from the one put forward by Bell and Barbour in (Chap. 3). To introduce this notion, I am going to discuss the viability of my approach for solving what I call the discontinuity problem of Boltzmannian statistical mechanics (BSM) and its import on the question of the reducibility of thermodynamics to BSM in (Sect. 3.1). Following the usual presentation of the framework (Sect. 3.1.1), the available phase space of a system is partitioned into a finite set of disjoint macro regions, corresponding to different macro states (Sect. 3.1.2). This construal of macro regions leads to the discontinuity problem of BSM: upon the micro state evolving from one macro region to another, it crosses boundaries between them, thereby instantiating different macro values associated with these macro regions. As a result, the usual framework of BSM invariably models change of macro values, whenever they change at all, as a jump discontinuity (Sect. 3.1.3). I argue that this is especially problematic for those who take BSM to be a fundamental theory (Sect. 3.1.4). The discontinuity problem has not been made explicit in the literature so far, although there are some hints to be found, to the effect that the usual way of partitioning phase space into disjoint sets with sharp boundaries is not considered entirely unproblematic by the community (Sect. 3.1.5). Hence, I will present its consequences in a series of examples (Sects. 3.1.6–3.1.8). I will consider two seemingly obvious approaches to tackle the problem, removing the underlying assumptions of the standard framework. As we will see, none leads to acceptable results (Sect. 3.1.9). In (Sects. 3.1.10–3.1.12), I show how the discontinuity problem poses a threat to the foundational project of reducing thermodynamics to BSM. I will then introduce a new framework for BSM (Sect. 3.2) that allows carving up phase space, not into disjoint macro regions, but potentially overlapping worlds, defined as sets of phase points sufficiently similar to a given micro state, i.e., instantiating macro states indistinguishable from a given macro state. Worlds thus construed, other than macro regions, can overlap consistently. As a result, the need to model the change of macro values as discontinuous jumps can be avoided. I argue that the spirit and basic features of BSM remain intact despite replacing macro regions with worlds. In particular, worlds, like macro regions, can be compared with other worlds (and the entirety of available phase space) in terms of their volumes, thus allowing for similar accounts of equilibrium states and approach to equilibrium as usual.

In the next chapter (Chap. 4), I will make use of this notion of worlds to solve a similar discontinuity problem as it arises in TBM. There, it is put into sharp relief: not only do remote configurations instantiate different macro states, but individual configurations aren't linked up by continuous trajectories. That is, not even on the level of the microscopic description must a continuous evolution be assumed. This lack of continuity in the micro evolution of the universe, following Bell, leads to the problem of temporal solipsism. To circumvent it, one could simply adopt the Everettian notion of worlds, understood as entire, decoherence-induced world branches. In particular, if TBM should turn out to be many worlds in denial, such that the particle configuration turns out to be superfluous theoretical structure—as, according to Brown, is the case in BM—this move would be appropriate. However, TBM features fundamental particles. At least in this respect, it is a bona fide Bohmian theory. In Bohmian theories, the natural basis for construing worlds are particle configurations, not the WF (Sect. 4.1). Also, Brown's criticism of standard BM doesn't carry over to TBM. The latter is not many worlds in denial; its particles are not superfluous theoretical structure. Rather, they are the substance material stuff is made of (Sect. 4.1.1). With particles playing the role of a substance in TBM's ontology, an account of psychophysical parallelism should take them to be the physical aspect of the psychophysical. Thus, worlds should be construed from particle configurations (Sect. 4.1.2). Accordingly, I will introduce the notion of worlds as sets of particle configurations into the framework of TBM (Sect. 4.2). For this, I make use of a nontransitive relation—sufficient similarity—already familiar from the previous chapter on BSM. In order to make my approach work, I discuss the issue of superpositions in TBM (Sect. 4.3), and the import of the microscopic-macroscopic distinction on the construal of worlds (Sect. 4.4). I will argue that Barbour's time-capsule approach is not suitable for the purpose of construing worlds, and ultimately, world histories (Sect. 4.5). It is deficient, in particular, when taking into consideration the background assumption—a more than minimal version of psychophysical parallelism—he makes, and which I share: for the impression of macroscopic change to emerge, Barbour, in accordance with our current best theories, relies on brain processes to happen. However, brain processes cannot be had from within individual, static configurations. While not logically being ruled out, time capsules alone don't seem to provide an adequate solution to temporal solipsism. Accordingly, worlds should not be identified with Barbourian time capsules, but with sets of configurations that are sufficiently similar in the sense of not amounting to macroscopic differences for a given macroscopic object within a configuration (Sect. 4.6). Before finally introducing worlds capable of forming FAPP-continuous world histories into the framework, I examine some preliminaries regarding the entities worlds are supposed to consist of, and the space they inhabit

(Sect. 4.7). In particular, issues revolving around particle identity, impenetrability and particle number conservation will be discussed (Sect. 4.7.1). With the appropriate notion of worlds thus developed (Sect. 4.8), world histories, FAPP-continuously linking up remote worlds in configuration space and time, can emerge (Sect. 4.9), thereby solving the issue of temporal solipsism while being compatible with a more than minimal psychophysical parallelism (Sect. 4.10).

Reference

Dürr, P., & Ehmann, A. (2021). The physics and metaphysics of Tychistic Bohmian mechanics. *Studies in History and Philosophy of Science Part A, 90*, 168–183. https://doi.org/10.1016/j.shpsa.2021.09.014

Chapter 2
A Short Introduction to Tychistic Bohmian Mechanics

To start off with a very quick overview, let me state the core features of Tychistic Bohmian Mechanics. TBM can be considered a descendant of Bohmian Mechanics and the Everett interpretation (henceforth "Everett" for short) of non-relativistic Quantum Mechanics (QM). It is a Bohmian theory in the sense that it features a similar ontology as BM, namely particles endowed with definite locations in physical space at all times. These particles in 3-space are regarded as fundamental in a sense to be made explicit. Nevertheless, TBM does not necessarily fall under the primitive ontology approach, although it can be understood as a primitive ontology theory, as we will see shortly.[1] For convenience, I will often refer to the particle configuration represented by a point in 3 N-dimensional configuration space, sometimes called the super-particle,[2] rather than to the N-particle system in 3-dimensional physical space. This does not imply some kind of ontological order between the two: The N-particle system in 3-dimensional physical space and the 1-super-particle system in 3N-dimensional configuration space are one and the same thing, merely represented differently. The super-particle, i.e., the configuration of the N-particle system as a whole, is actualized at a single point in configuration space. TBM deviates from orthodox BM in that neither particles in 3-space nor the super-particle move on continuous trajectories. Instead, configurations are actualized at random, following the quantum equilibrium hypothesis (QEH), understood in TBM as a physical law stochastically restricting the actualizations of the super-particle in accordance with the probabilities given by the Born rule (BR): The super-particle jumps through configuration space, following the probability density associated

[1] The primitive ontology approach shall be explained in II.4. For reasons to be elaborated there, I am hesitant calling TBM's ontology a primitive ontology sensu stricto without further qualification.

[2] Or "universal particle" or "world-particle" in Albert's terminology, cf. Albert (1996, p. 278).

A. Ehmann, *Worlds in a Stochastic Universe: On the Emergence of World Histories in Tychistic Bohmian Mechanics*, Synthese Library 537,
https://doi.org/10.1007/978-3-032-14632-8_2

with the universal wave function (WF). The WF itself, as usually, evolves deterministically according to the standard Schrödinger equation (SE).

From these jumps of the super-particle through configuration space, together with the branching structure of the WF induced by decoherence, stems the main feature TBM inherits from Everett, namely its many-worlds character: at different times, the super-particle actualizes different configurations and, a fortiori, instantiates different world branches. The disposition for a given configuration to be actualized is encoded in the WF, with its probability—understood as a (single-case) propensity—given by the BR. TBM's actualizations are genuinely, irreducibly chancy, but weighted: individual actualizations are always probabilistically restricted by the QEH in such a way that they collectively approximate Born statistics.

A more detailed picture of TBM will emerge from a comparison with standard BM. Alterations of BM's formalism, and a reinterpretation of probabilities in terms of propensities, as just mentioned, lead to the formulation of TBM. TBM turns out to be empirically equivalent with orthodox QM, as required of every viable alternative theory. Before giving a quick outline of standard BM, however, a few remarks about terminology are in order:

It is important to be clear about which theory we refer to when speaking of (standard) Bohmian Mechanics. In particular, Bohmian Mechanics as put forward by Dürr et al. (1992), despite its name, must not be confused with Bohm's original theory as set out in his 1952 papers (Bohm, 1952a, 1952b). The latter goes under different names. Sometimes it is called "the Bohm interpretation of quantum mechanics", or simply "the Bohm interpretation" (Hiley, 2009, passim), sometimes "causal" or "ontological interpretation" (cf. Friebe et al., 2015, p. 195). Goldstein (2017, §5) refers to it as "the quantum potential formulation of the de Broglie-Bohm theory" or "the quantum potential formulation of Bohmian mechanics". Cushing heads Chap. 4 of his 1994 book "Bohm's Quantum Theory", but usually refers to it simply as "Bohm's theory" (Cushing, 1994, passim).[3] I shall follow Cushing in calling this theory "Bohm's theory", or alternatively "Bohm's original theory", where the emphasis is called for. All these names refer to a theory that is different from what I, in accordance with Dürr et al. (1992), call Bohmian Mechanics, both formally and ontologically. Cushing (1994, §2.1, 9, emphasis in original) remarks that "a scientific theory can be seen as having two distinct components: its formalism and its interpretation", where "formalism means a set of equations and a set of calculational rules for making predictions that can be compared with experiment", and "interpretation refers to what the theory tells us about the underlying structure of these phenomena (i.e., the corresponding story about the furniture of the world – an ontology). Hence, *one* formalism with *two* different interpretations counts as *two* different theories". Adopting this distinction, Bohm's original theory and Bohmian Mechanics turn out to be decidedly different theories, for they differ in both

[3] However, he also refers to it (ibid., 47) as "the causal interpretation".

formalism and interpretation. The former involves a quantum potential[4] that contributes to a force acting on the particles (Bohm, 1952a, p. 170): "the particles moves [sic] under the action of a force which is not entirely derivable from the classical potential [...] but which also obtains a contribution from the 'quantum-mechanical' potential". In Bohmian Mechanics, by contrast, the quantum potential plays no fundamental role. There are no fundamental forces acting on Bohmian particles (Goldstein, 2017, §5): "Bohmian mechanics as presented here is a first-order theory, in which it is the velocity, the rate of change of position, that is fundamental. It is this quantity, given by the guiding equation, that the theory specifies directly and simply. The second-order (Newtonian) concepts of acceleration and force, work and energy do not play any fundamental role. Bohm, however, did not regard his theory in this way. He regarded it, fundamentally, as a second-order theory, describing particles moving under the influence of forces, among which, however, is a force stemming from a 'quantum potential'." Considering their different interpretations, i.e., their different ontologies, it becomes obvious that Bohm's original theory and Bohmian Mechanics are not the same theory. The former, with its quantum potential generating a force on particles, provides "a story about the furniture of the world" quite different from the story told by the latter. As reported by Hiley (1999, p. 117), Bohm agrees with this assessment: "It should be noted that the views expressed in our book [Bohm & Hiley, 1993] differ very substantially from those of Dürr et al. (1992) who have developed an alternative theory. It was very unfortunately that they chose the term 'Bohmian mechanics' to describe their work. When Bohm first saw the term he remarked, "Why do they call it 'Bohmian mechanics'? Have they not understood a thing that I have written?" [...] It would have been far better if Dürr et al. (1992) had chosen the term 'Bell mechanics'. That would have reflected the actual situation far more accurately." Indeed, a lot of terminological confusion could have been avoided. However, that horse has left the barn. Goldstein et al.'s theory is standardly referred to as Bohmian Mechanics, and I will adhere to that terminology.

2.1 Bohmian Mechanics as a Solution to the Measurement Problem

BM sets out to complete the formalism of QM and thus to provide a solution to what is generally referred to as "the measurement problem" of QM, arising from the general presence of superposition states.[5] Maudlin (1995) provides a concise statement of, as he calls it (ibid., 7), the "problem of outcomes". He presents three

[4] And restrictions on particle momenta to render the theory empirically equivalent with QM, see Bohm (1952a, p. 170), Cushing (1994, p. 43 and 45).

[5] This is not to say that the solution of the measurement problem was the motivation behind the development of BM from the historical beginning, but merely that it is one of the main motivations for adopting BM today.

propositions at most two of which can be true at the same time. The fact that the conjunction of these propositions is inconsistent constitutes the measurement problem. Paraphrasing Maudlin:

(A) The WF provides a complete description of the physical properties of a system.
(B) The WF always evolves according to the SE.
(C) Measurement outcomes are always determinate.

To see why (A), (B) and (C) can't all be true, think of the following example of the measurement of the spin of an electron, in a certain direction. The outcome of such a measurement is determinate if (ibid.) "at the end of the measurement the measuring device is either in a state which indicates spin up (and not spin down) or spin down (and not spin up)" in that direction.

Before the measurement, the device is in its pre-measurement ready state, $|ready\rangle_d$.[6] Let's say it is a device that measures spin in z-direction, and we prepare an electron with spin up in that direction, written as $|\uparrow_z\rangle_e$. The expression $|\uparrow_z\rangle_e \otimes |ready\rangle_d$ denotes the state of the total system, consisting of the electron and the measuring device, before the measurement. After the measurement, the state of the total system will be $|\uparrow_z\rangle_e \otimes |up\rangle_d$, with $|up\rangle_d$ indicating that the measuring device shows the outcome "up", as expected of a properly working device.

If we prepare the electron to be in the state $|\uparrow_x\rangle_e$ (spin up in x-direction), say, it will be in a superposition of z-spin states, which can be expressed as $|\uparrow_x\rangle_e = 1/\sqrt{2}$ $|\uparrow_z\rangle_e + 1/\sqrt{2}$ $|\downarrow_z\rangle_e$. Accordingly, the pre-measurement state of the total system is

$$|\uparrow_x\rangle_e \otimes |ready\rangle_d = \left(1/\sqrt{2}|\uparrow_z\rangle_e + 1/\sqrt{2}|\downarrow_z\rangle_e\right) \otimes |ready\rangle_d$$

Proposition (B) above tells us that this pre-measurement state of the total system evolves into the post-measurement state $1/\sqrt{2}|\uparrow_z\rangle_e \otimes |up\rangle_d + 1/\sqrt{2}|\downarrow_z\rangle_e \otimes |down\rangle_d$. But being in this state, the measuring device isn't in a determinate state of either showing the outcome "up" or the outcome "down". It's in a superposition. According to (A), this is a complete description of the physical properties of the system. So, assuming that (A) and (B) are true, the evolution of the system according to the SE results in the system being in an indeterminate state, and this state is the complete description. This contradicts proposition (C).

Ways out of this predicament are to abandon one of the three propositions. Everettians will deny that (C) is true. Proponents of collapse[7] theories reject (B). Bohmians hold that the WF isn't a complete description of the physical properties of a system. They thus abandon (A) and attempt to complete the description with a definite particle configuration.

---

[6] Except for some adaptations, I follow Maudlin's example and notation.

[7] As Maudlin notes (ibid, 8), it would be better to speak of non-linear evolution instead of collapse, since there are theories in which the evolution of the WF is always non-linear. However, so-called collapse interpretations are the most prominent amongst those featuring a non-linear evolution of the WF.

BM, with its completion of the state description by adding a definite particle configuration, indeed can provide a solution to the measurement problem. In light of the severity of this problem, this provides good motivation to consider Bohmian theories as viable alternatives to orthodox QM and its various interpretations. However, for a theory, or a class of theories, to be viable alternatives to a given theory, it is not sufficient to solve some problem of the latter, be that as important as it may. Being a viable alternative requires empirical equivalence with the theory to be replaced. In the following sections, I will therefore first provide a general description of BM and establish its empirical equivalence with orthodox QM, and then give an account of TBM's empirical equivalence with BM. Via the transitivity of empirical equivalence, TBM can be said to be empirically equivalent with QM, iff it is empirically equivalent with BM, and BM is empirically equivalent with QM. So let's first describe BM and its empirical equivalence with QM in more general terms.

2.2 A General Description of BM and Its Empirical Equivalence with QM

BM is a non-relativistic quantum theory. It describes a universe consisting of N particles and their deterministic, continuous evolution in space. The particle configuration Q_t at time t is given by the definite particle positions of the N particles at that time. BM adheres to the standard, non-relativistic SE, describing the evolution of the universal quantum wave function Ψ for a universe of N particles. The guiding equation (GE), featuring Ψ, describes the evolution of the universal particle configuration Q in terms of continuous trajectories of the Bohmian particles. Given some initial configuration Q_{t_0}, i.e., the initial spatial positions of all N particles at initial time t_0, the GE determines the configuration for all times t_i, before and after t_0. The quantum equilibrium hypothesis (QEH) states that (initial) ensembles of identically prepared subsystems of the universe, each with the WF ψ, typically, i.e., almost always, are in fact Born distributed with theoretical distribution $\rho_{th} = |\psi|^2$, approximating the empirical distribution ρ_{emp} for large ensembles of identically prepared subsystems.[8] This requires that ours is a typical universe, and that in typical universes, typical ensembles are Born distributed.

Justifying the typicality approach to BM is a rather intricate matter. I am unable to engage with it here. However, the alleged typicality of our universe notwithstanding, since BM is a deterministic theory and contains the assumption that particle positions are always definite, its probabilities for finding an individual particle at a definite location are often construed as a measure of the ignorance of an observer

[8] Cf. Goldstein (2012, p. 66): "The theoretical distribution is an idealization providing a good approximation to the empirical distribution, $\rho_{emp} \approx \rho_{th}$, in the limit of large ensembles of subsystems."

about the precise locations of particles, especially their initial distribution.[9] While the typicality approach explains why one finds the distribution of particles in *ensembles* of identically prepared systems as $|\psi(Q)|^2$, it does not do away with the ignorance interpretation of probabilities for the position of particles in *individual* systems. For them, an observer can only provide probabilities to be here or there, because they don't *know* where the particles *actually* are. But since, according to BM, they must be at a definite location, which evolves deterministically according to the GE, these probabilities do not reflect a fundamentally stochastic nature, but the observer's ignorance.[10]

Especially if one regards BM to be a "quantum theory without observers" (cf. Allori et al., 2008, passim), it prima facie seems unclear whose ignorance about the initial configuration it is that is reflected in BM's QEH. Also, shouldn't probabilities be construed as fundamental and objective in a theory that features them at its heart and claims to be objective, without observers? My intuition tells me that they should. But irrespective of whether your intuition agrees with mine on that matter, on TBM's irreducibly stochastic actualization dynamics, probabilities are construed as fundamental and objective. If you share my intuition, you might regard this as an advantage TBM has over BM.

QEH and the GE together guarantee empirical equivalence between BM and QM. For example, the particle positions measured at the screen of a repeated double slit experiment will (almost certainly) approximate the distribution expected according to BR. The GE ensures that this is always the case: *If* an ensemble of initial configurations already approximates the Born distribution, it will also approximate the Born distribution at earlier and later times, because the dynamics given by the GE preserves the initial distribution. Hence, in the experiment just mentioned, if the ensemble was (approximately) Born distributed at its preparation, it will be at the time of measurement at the screen, and vice versa. In other words, regarding

---

[9] See, e.g., Bell (1987b, p. 112): "the only use of probability here is, as in classical statistical mechanics, to take account of uncertainty in initial conditions." Likewise, Passon (2010, p. 53): "The result of the experiment is random only for us, because we don't know the initial conditions." (My translation.) Callender (2007, p. 364) acknowledges this fact—"When speaking of probabilities in a deterministic theory one often hears it said that the probabilities are 'epistemic,' i.e., due to ignorance"—but rejects the ignorance interpretation.

[10] With respect to this ignorance interpretation of probabilities, BM resembles Bohm's original theory, (cf. Bohm, 1952a, p. 166): "In contrast to the usual interpretation, this alternative interpretation permits us to conceive of each individual system as being in a precisely definable state, whose changes with time are determined by definite laws, analogous to (but not identical with) the classical equations of motion. Quantum-mechanical probabilities are regarded (like their counterparts in classical statistical mechanics) as only a practical necessity and not as a manifestation of an inherent lack of complete determination in the properties of matter at the quantum level." As the reason for this practical necessity, he cites "uncontrollable disturbances" (ibid., 171): "In the usual interpretation, however, the need for a probability description is regarded as inherent in the very structure of matter […], whereas in our interpretation, it arises […] because from one measurement to the next, we cannot in practice predict or control the precise location of a particle, as a result of corresponding unpredictable and uncontrollable disturbances introduced by the measuring apparatus."

observable phenomena, BM comprises the same claims as QM does: the theories are empirically equivalent.

Apart from their positions, the usual dynamical properties of particles—the properties that can change over time—are contextual in BM. Spin is an example of such a contextual property of a particle, a quantity that supervenes on the complete state description WF + configuration. It is a way of describing the particle's behaviour in certain contexts. Accordingly, the spin of a particle is determined by its behaviour under certain experimental conditions, e.g., in a Stern-Gerlach experiment. Recall that also in orthodox QM, a spin measurement ultimately is a position measurement: one of two detectors registers an event, while the other doesn't. Whether this is interpreted as a Bohmian particle ending up in said detector, or as a wave function collapsing onto the corresponding eigenstate upon measurement, makes a difference in metaphysics and ontology, but, empirically, amounts to the same phenomenon. Again, the theories exhibit empirical equivalence.

So much for this quick outline of BM and its empirical equivalence with QM. It shall provide the background against which, in the following section, TBM is discussed. In particular, I will establish TBM's empirical equivalence with BM, and a fortiori with QM.

2.3 From Bohmian Mechanics to Tychistic Bohmian Mechanics

In essence, TBM is the somewhat elaborate claim that the GE is not strictly necessary for a Bohmian theory, i.e., that a Bohmian theory without a GE is viable, in the sense that it is internally consistent and empirically equivalent to standard BM—and hence orthodox QM—and metaphysically adequate.

To be more precise, TBM, formally, retains the SE (including the WF) and QEH from BM, but not the GE. In addition, it reinterprets the QEH as a stochastic law for the actualization of configurations. Together, the SE, the WF and the QEH guarantee empirical equivalence: the SE provides the dynamics for the WF, and the WF as featuring in the QEH restricts the distribution of particles in ensembles, in the same way it did in BM for the so-called "initial" distribution. But while in BM, the role of the QEH is most obvious for this initial distribution, TBM makes its role explicit for all distributions, at all times. Recall that, in BM, particle positions are evolved by the GE in such a manner that the distribution is preserved. If an ensemble is Born distributed once, it is for all times.[11] But note that, under the condition that QEH is

[11] Not taking into account the possible relaxation of non-equilibrium probability distributions into equilibrium as shown by Valentini, which requires the assumption that the initial probability distribution is suitably coarse-grained (Valentini, 1991, p. 8): "The proof rests on the assumption of no 'micro-structure' for the initial state, as for the classical case. Specifically, we assume the equality of coarse-grained and fine-grained quantities at the initial time t = 0, i.e., we assume $\bar{P}(0)=P(0),\overline{|\psi(0)|^2}=|\psi(Q)|^2$." As Norsen (2018) shows, this allows for the evolution of the

assumed for all times instead of merely some initial time t_0, the WF, SE and QEH alone, i.e., without the GE, provide sufficient theoretical structure to guarantee Born distribution at all times. More colloquially: If we feel justified to assume QEH for some *arbitrary* point in time, why shouldn't we feel justified to assume it for *any* point in time, calling the distribution at this time the "initial" distribution, if so desired? Of any distribution, QEH will tell us that it approximates Born statistics. One might be compelled to reply that it is simpler to assume QEH just for one instant and let the dynamics do the work. But note that this requires an additional assumption itself, namely the deterministic particle dynamics as represented by the GE. In effect, we are replacing two assumptions—that QEH holds for the initial configuration, and that the GE evolves the configuration from there on, such that the ensemble always approximates Born distribution—by one assumption, namely that the QEH holds for all configurations, at all times. Arguably, the exchange is worthwhile: While avoiding the assumption of a suitable GE, including the metaphysical baggage this brings with it, we merely extend the assumption that the QEH holds for (at least) one instant to the assumption that it holds for all instants. As long as there is no strong reason against it, this seems fair. After all, standard BM includes the same claim of the ensemble always being (approximately) Born distributed, if it ever is. The difference between BM and TBM is how we arrive there: TBM assumes applicability of the QEH at all times. BM arrives there via the assumption of the applicability of the QEH at some (initial) time *plus* the assumption of an appropriate particle dynamics (GE) that guarantees Born distribution at all other times as well, given Born distribution at the initial time.

Let's elaborate. Does the GE play an indispensable role in the preservation of Born distribution over time? According to TBM, it doesn't. In BM, the GE serves a dual purpose: The first is to evolve the positions of individual particles *deterministically*. The second is to do this in such a manner that the ensemble distribution at all times is compatible with the restrictions imposed by QEH, given the initial distribution is. QEH in BM is exactly that, "the assumption that the initial (t = 0) configuration of a quantum system was random with distribution P_B [i.e., Born distribution]" (Norsen, 2018, p. 5). Regarding its first purpose, it is easy to dispense with it, if one doesn't insist on determinism.[12] Regarding its second purpose, within the

probability distribution not only from non-equilibrium to equilibrium, but also from equilibrium to non-equilibrium. Given we do not assume QEH and instead follow Valentini's derivation, it cannot be claimed that an ensemble is Born distributed at all times, if it is Born distributed once. However, for TBM, we assume QEH, such that the claim still holds.

[12] Even proponents of BM, despite regarding BM as (Goldstein et al., 2011, p. 12) "the simplest satisfactory version of quantum mechanics", don't feel particularly committed to determinism: "The circumstance that it uses point particles and is deterministic does not mean that we are dogmatically committed to point particles or determinism; also indeterministic theories with another ontology may very well be satisfactory." In the same spirit, they (Dürr & Teufel, 2009, p. 9) reject the statement that "the aim of Bohmian mechanics is to restore determinism in the quantum world. That is false. [...] What is 'out there' could just as well be governed by stochastic laws [...]. It happens to be deterministic, which is fine, but not an ontological necessity."

framework of TBM, the GE is unnecessary as well. Taking the QEH as restricting not only the *initial* configuration, but the random actualizations of configurations in general, ensembles of actualized configurations will (approximately) show the required Born distribution. Postulating trajectories, then, is superfluous. Following Norsen's (2018) presentation of equivariance, one could say that with the equivariance of $\rho = |\Psi|^2$, "the usual quantum 'probability density'" (ibid., 4) or "wave function 'intensity'" (ibid., 5), the equivariance of the "particle probability distribution" P_B in TBM is ensured, quite trivially, via the QEH. Additional dynamics preserving the initial distribution over time isn't necessary. Either way, equivariance is achieved, and it is equivariance that is responsible for the fact that the actual, empirical particle distribution "tracks" the probability density. Because both theories have this feature, they are empirically equivalent, and a fortiori, because of the empirical equivalence of BM and QM, TBM and QM are empirically equivalent as well. In conclusion, for empirical equivalence with QM, trajectories are not needed.

This argument by itself might not convince everyone to adopt TBM over standard BM. One might have extra-empirical reasons to prefer a theory with continuous, deterministic trajectories, despite them not being strictly necessary. In particular, one might argue with Bell that a fundamentally stochastic dynamics of the configuration leads to a temporal form of solipsism. This issue will be discussed at length. For now, I am not going to take sides on the matter. My present claim is merely that TBM is a viable alternative to BM, given one doesn't insist on a deterministic, continuous particle dynamics, because empirically, the two theories are equivalent.

How, then, do particle configurations in TBM evolve in such a manner that QEH holds at all times? This is due to QEH restricting the random, discontinuous particle dynamics that replaces the GE. In TBM, the particle configuration—i.e., the super-particle—of the universe is assumed to instantaneously jump through 3N-dimensional configuration space, in accordance with QEH. Accordingly, the N Bohmian particles locate and dislocate themselves in 3-dimensional physical space. Without requiring any external triggering mechanism—after all, the universe by definition is all there is, such that there is nothing external to it—these jumps of the super-particle are spontaneous. The disposition for certain configurations to be actualized is encoded in the WF, and the respective probabilities (propensities) as given by BR—being a measure of a fundamental feature of the universe, namely said disposition—are objective. In contradistinction to BM's deterministic particle dynamics, actualizations in TBM at any time are fundamentally random. They are stochastically independent from each other: $P(Q_t \cap Q_{t'}) = P(Q_t) * P(Q_{t'})$.[13] There is no

[13] Some remarks are in order regarding the stochastic independence of TBM's actualizations at different times. At first blush, the stochastic independence of actualizations may seem counterintuitive. Underlying this impression is the following thought: the probability density distribution over configuration space is determined by the WF. The latter evolves deterministically, governed by the SE, so the probability density at one time isn't independent of the probability density at another time. The actualization at one time might be informative about the probability density at that time, and via the latter's deterministic evolution, about the probability density at other times.

link between actualized configurations at different times. In Bell's (1987a, p. 135) words: "there is no association of the particular present with any particular past." But as has been argued, this link, which would be provided by trajectories in BM, is not necessary for empirical equivalence with orthodox QM. The stochastic dynamics of the super-particle, following the probability density as it jumps, is sufficient, because it adheres to the restrictions imposed by QEH, and being in accord with QEH at all times is sufficient for being empirically equivalent with QM.

Having established the empirical equivalence of TBM, BM and QM, despite the former's lack of continuous trajectories, I shall now turn to two interrelated issues: TBM's classification as a primitive ontology theory, and its solution to the problem of communication, arising from the two-space reading of BM. Many proponents of BM deem a theory's classification as a PO crucial for it being bona fide Bohmian. In light of this, we must ask whether TBM can be made compatible with the PO approach. After presenting the PO approach, I will argue that this is indeed possible, but that, in order to also solve the problem of communication, our interpretational freedom with respect to the WF is restricted. In particular, subsuming TBM under the PO paradigm seems to require abandoning the view that the WF encodes a disposition.

2.4 TBM as a PO Theory and the Problem of Communication

As just noted, TBM *can* be understood as a primitive ontology theory, when imposing certain restrictions on the interpretation of the WF. In principle, at least in my view, nothing much hinges on this classification. However, since proponents of BM often regard applicability of the PO framework as crucial for a theory to be Bohmian, I want to show that TBM can be made compatible with it, thereby pre-empting the criticism that TBM is not a viable alternative to BM because it cannot be classified as a PO theory. Abandoning an ontologically thick, dispositionalist view of the WF (and instead adopting an ontologically thin stance towards the WF) opens up this possibility.

Thus, one might be tempted to conclude, the actualization of $Q_{t'}$ at t' isn't stochastically independent of the actualization of Q_t at t. This reasoning is misleading, however. In responding as follows, I hope to make clear that TBM's actualizations indeed are stochastically independent.

On TBM, there is no collapse of the universal WF upon the actualization of a configuration. If there was, an actualization would suddenly change the WF and hence the probability density distribution for subsequent actualizations. Because there is no such collapse, the probability density distribution governing the actualization of $Q_{t'}$ at t' isn't influenced by the actualization of Q_t at t. We may therefore take the probability of the joint event to factorise. The stochastic independence of actualizations is not affected by the lawlike dependence between probability density distributions at different times. The latter are related to each other via the deterministic evolution of the WF, governed by the SE. But this has no bearing on the fact that actualizations at different times don't affect each other.

I thank Jiji Zhang (Lingnan) for pressing me on clarifying this point.

2.4.1 *What Makes a PO Theory?*

To tackle this issue, we first need to be clear about the central features of the PO approach. What makes a PO theory? Following the literature, the PO approach imposes a strict distinction between primitive and non-primitive elements of the ontology of a theory. It is only the primitive elements that are fundamental and make up/compose non-fundamental physical objects. Allori (2015, p. 107f.) writes: "what constitutes physical objects, the PO, [...] lives in three-dimensional space or space-time and constitutes the building blocks of everything else. In the formalism of the theory, the variables representing the PO are called the primitive variables. In addition, there are other [non-primitive] variables necessary to implement the dynamics for the primitive variables". Pertaining to fundamental theories, the primitive variables (ibid., 109) "are ontologically primitive, given that they represent matter and they provide the fundamental entities the theory describes."

In order to make the notion of PO more precise, this fundamentality claim should be spelled out. It consists of two separate claims: (1) The PO is fundamental in the sense that all other physical objects are composed of its elements. (2) The PO is fundamental in the sense that it itself is not ontologically dependent on further, non-3D entities.

I shall refer to the kind of fundamentality claimed in (1) as $\text{fundamental}_{\text{CMB}}$: The primitive entities are $\text{fundamental}_{\text{CMB}}$ in that they form the *complete minimal basis* (CMB) for composed physical objects.[14] That is to say, the PO contains all and only those 3D entities from which all other physical objects are made. The qualifications "complete" and "minimal" are of some importance. The former ("complete") specifies the PO such that it contains *all* the elements necessary to make up other physical objects. The reason for imposing this completeness requirement should be immediately obvious: if the PO were not complete, not all other physical objects could be composed from its elements. The latter ("minimal") specifies the PO such that it contains *only those* elements. I.e., the PO does not contain entities beyond those necessary to compose all other physical objects. The reason for imposing this requirement is fairly obvious, too: Without insisting on the PO being minimal in order to be fundamental, we could simply regard the set of all physical objects, including composed objects, to be the PO.[15] With the PO being fundamental, this would render composed objects fundamental as well. At best, this amounts to an empty notion of fundamentality. At worst, it is an outright contradiction. In any case, it is at odds with the PO approach.

---

[14] I thank an anonymous reviewer of my paper with Patrick Dürr for pressing on this further specification of the PO in terms of a complete minimal basis.

[15] Cf. Tahko (2018, 1.3): "We have included a *minimality* condition in the definition of (CMB). This is an important addition, because otherwise we could take the set of all the entities in the world and call it complete, given that this set as well would include all the fundamental entities. So, according to (CMB), the complete minimal basis must include all and only the fundamental entities."

The kind of fundamentality claimed in (2) shall be referred to as fundamental$_{\mathrm{ODEP}}$. Since it contains the notion of ontological dependence on non-3D entities, we must flesh out the meaning of this notion. Oldofredi & Esfeld (2018, p. 11) demand that the elements of the PO be fundamental (i.e., fundamental$_{\mathrm{ODEP}}$) in the sense that "they are not reducible to more elementary notions". Ontological independence, then, simply means to be irreducible in this sense. However, this seems to be not precise enough, for ontological independence in terms of irreducibility can mean several things. For example, an entity can be said to be irreducible if it isn't instantiated by a non-3D entity playing its functional role,[16] or if it isn't identical to a non-3D entity.[17]

Demanding irreducibility only with respect to *non-3D* entities requires an explanation. Arguably, it already would be a blow to a PO theory if its PO should turn out to be reducible to *3D* entities, and hence to "more elementary notions". But that blow wouldn't be fatal. It would still be possible to revise the theory, such that it contains as its new PO those 3D entities to which its former PO turned out to be reducible. As long as the new PO is not reducible any further, it can be regarded as fundamental$_{\mathrm{ODEP}}$. The revised theory, then, remains a PO theory. Reducibility to 3D entities thus might be an inconvenience, but it doesn't hinder the PO approach in principle. However, reducibility of the PO to non-3D entities is fatal to the PO approach. In this scenario, the theory can't simply be revised such that it contains as its new PO those entities that turned out to be fundamental, because the PO approach doesn't allow for non-3D entities in the PO. Hence, demanding irreducibility only with respect to non-3D entities seems apt: the PO must be ontologically independent in that it is not reducible to non-3D entities, e.g., the 3N-dimensional WF. That is, no non-3D entity may be playing the PO's functional role or be identical with it.

In light of this restricted demand, there is another type of ontological dependence that should be mentioned, because it plays a somewhat special role in the context of classifying theories as PO theories. One could argue that entities are not fundamental$_{\mathrm{ODEP}}$ if they have proper mereological parts. That is to say, fundamental$_{\mathrm{ODEP}}$ entities should be mereologically simple. At first blush, a PO not being mereologically simple seems to be fixable just as easily as general dependence on 3D entities: if elements of the PO turn out to have proper mereological parts, the theory can be revised, such that the new PO consists of those proper mereological parts. The PO

[16] On Everett, particles, just like macroscopic objects, are reducible in this sense: They are structural patterns, instantiated by the (non-3D) universal WF, playing the functional role of such objects. As such, they are useful ways of describing reality, albeit at a non-fundamental level (cf. Wallace, 2012, Ch. 2). Fundamentally, all there is is the universal WF.

[17] In addition to these options, one might be tempted to spell out reducibility in terms of supervenience. An entity, then, would be reducible if it supervenes on a non-3D entity, i.e., if change in the former cannot be had without change in the latter. However, while supervenience seems necessary for reducibility, it is unclear whether supervenience suffices for reducibility, such that the latter could be spelled out in terms of the former. As McLaughlin & Bennett (2018, 3.3) put it: "Everyone agrees that reduction requires supervenience. […] The more interesting issue is whether supervenience suffices for reduction". It lies outside the scope of this book to engage with this debate. I will therefore not further discuss supervenience as a candidate sufficient for spelling out reducibility of the PO. However, I will say more about supervenience and reduction in III.

still wouldn't be reducible to non-3D entities. However, an issue arises if there can't be mereological simples in the PO, such that it can never be fundamental. Consider the GRW matter density theory (GRWm), usually classified as a PO theory: with the matter density being continuously spread over physical space, every mereological part of it has further, proper mereological parts. As Esfeld (2018, p. 174) diagnoses, "the matter density stuff, being gunk, is infinitely divisible". Being infinitely divisible, the matter density cannot be fundamental$_{\text{ODEP}}$, not because it itself has proper mereological parts, which would be fixable by revising the theory such that the parts are regarded as the PO, but because its proper mereological parts have proper mereological parts as well, and so on, ad infinitum. Thus, it is impossible to arrive at a revision featuring a truly fundamental$_{\text{ODEP}}$ PO. To resolve this issue, we can either decide to not regard GRWm as a PO theory, or to allow for fundamental$_{\text{ODEP}}$ entities not being mereologically simple. I leave this for the proponents of the PO approach to decide. Since TBM's particles aren't infinitely divisible, this is not an issue for the present investigation.

To sum up: the primitive ontology of a theory consists of those and only those 3D entities that are fundamental$_{\text{CMB}}$—i.e., that form a complete minimal basis—to all other physical objects, and that are fundamental$_{\text{ODEP}}$, i.e., that are not reducible to other, non-3D entities, in the sense elaborated.

2.4.2 Is TBM Compatible with the PO Approach?

Prima facie, TBM's particles aren't reducible to non-3D entities: they are not instantiated by non-3D entities playing their functional role, aren't identical to non-3D entities, and don't have proper mereological parts. That is to say, they are fundamental$_{\text{ODEP}}$. In this respect, TBM seems compatible with the PO approach. At least for now.

But are TBM's particles fundamental$_{\text{CMB}}$ as well? This depends on the ontological status and role ascribed to TBM's WF. If the latter, under a given interpretation, turns out to be part of the complete minimal basis and hence of the primitive ontology, then, under this interpretation, TBM cannot be a PO theory, since the PO approach doesn't allow for non-3D entities in the PO, which the WF is.

An interpretation of the WF, however, must not simply be chosen at will, such that the theory can be classified as a PO theory. A further requirement is imposed, namely that the theory be able to solve the "problem of communication"[18] arising from the (Ney, 2012, p. 11) "two-space reading of Bohmian mechanics" (and likewise of TBM), i.e., arising from the fact that (ibid.) "the theory posits the existence of two physical spaces: a high-dimensional configuration space occupied by the wavefunction, and a separate, three-dimensional physical space occupied by the many particles that make up our tables, chairs, and the other objects of our manifest

[18] As it is called e.g. in Suárez (2015) and references therein.

image of the world." In short, to render TBM a PO theory, the interpretation of its WF must meet two demands at the same time: it must solve or circumvent the problem of communication while not considering the WF to be part of the complete minimal basis. This, I hold, can only be achieved under a "nomological" interpretation, i.e., when the WF is regarded as a law, or lawlike. To see this, let's have a look at the problem of communication.

2.4.3 The Problem of Communication

In standard BM, the N individual Bohmian particles, with position being their only non-contextual dynamical property, are located in ordinary, 3-dimensional physical space, and make up all physical objects, while the WF lives on 3N-dimensional configuration space, restricts their initial distribution and guides the evolution of their positions via the GE. This is the two-space reading of BM—particles living in 3-dimensional physical space, the WF living in 3N-dimensional configuration space—ultimately resulting in the problem of communication:

On the one hand, the WF's role in the theory suggests interpreting it as a real, physical object, e.g., a field. After all, how can an abstract, purely mathematical object, defined on likewise abstract and purely mathematical configuration space influence fundamental, real objects, i.e., particles, in physical 3-space? To explain this influence, the WF should be considered a real part of the complete, physical state description and its evolution, at least in the sense that it has an effect on the fundamental stuff (Ney, 2012, p. 11): "For even on this two-space reading of Bohmian mechanics, where the particles are ontologically fundamental, it is still true that it is states of the wavefunction that determine these particles' behavior over time. The particles themselves only have positions. And what they do at later times according to this theory, i.e., what positions they move to over time, depends on the wavefunction." Featuring in the GE, the WF guides the movement of particles on their trajectories through physical space (in a non-local manner). Likewise, it restricts their (initial) distribution, as it features in QEH. The central role the WF plays in the formalism, and for the empirical adequacy of BM, renders it difficult to immediately dismiss it as a purely mathematical artefact without any ontological significance. As Bell (1987a, p. 128, his emphasis) writes: "*No one can understand this theory until he is willing to think of* ψ *as a real, objective field rather than just a 'probability amplitude'. Even though it propagates not in 3-space but in 3N-space.*"[19]

On the other hand, how to account for the influence this real physical object has on the fundamental particles, given the different spaces they live in? How does a physical object in 3N-dimensional configuration space tell the stuff in 3-space what

[19] Note that this remark applies to BM as well as to Bohm's original theory with a quantum potential (cf. Suárez, 2015, p. 10f., as quoted in the following passage.) Bell himself, in his 1987a as well as in his unpublished paper from 1971 (see Sect. 2.5 below), refers to Bohm's work of 1952.

to do? This is the problem of communication. As Suárez (2015, p. 10f.) notes, "this is generally a problem for any account on which the wave-function in configuration space 'dictates' features of physical 3D space that are responsible for the motion of particles, whether it be a multifield, forces acting in the space, a quantum potential located in that space, or directly the motions of the 3D particles themselves. […] How does the wavefunction in configuration space *guide* the particles in physical space?"

There are several ways TBM can deal with this problem of communication. For example, one can regard the super-particle in configuration space as the fundamental$_{\mathrm{ODEP}}$ referent of TBM's ontology while not claiming such fundamentality for the Bohmian particles in 3-space. Such a move would immediately solve the problem (by circumventing it): adopting Albert's super-particle view, the fundamental$_{\mathrm{ODEP}}$ referent of the theory is put into the same space as the WF, i.e., 3N-dimensional configuration space. In contradistinction to the "two-space reading of Bohmian mechanics", this can be called the one-space reading of TBM. The super-particle actualizes instantaneous, definite configurations in configuration space, with the actualization disposition encoded in the WF, living on the very same space. The problem of an entity living in one space influencing another, living in a different space, no longer arises. At the same time, through the configuration being definite at all times, particle positions in 3-space are definite as well, allowing for the composition of physical objects. Bohmian particles in 3-space, hence, remain fundamental$_{\mathrm{CMB}}$. However, if we want TBM to be compatible with the PO approach as presented above, this option leads to a dead end: while, with adopting the super-particle as the fundamental$_{\mathrm{ODEP}}$ referent of the theory, Bohmian particles in 3-space remain fundamental$_{\mathrm{CMB}}$, they are no longer fundamental$_{\mathrm{ODEP}}$, for they either are secondary to the super-particle, or, at best, identical with it. Also, the super-particle being fundamental$_{\mathrm{ODEP}}$ conflicts with the PO approach, demanding that the PO should be entirely 3-dimensional. This can only be resolved if we think of the super-particle not as the fundamental referent of the theory, but as a mere representational variation, describing the configuration of the Bohmian particles in physical 3-space, such that the latter remain fundamental$_{\mathrm{ODEP}}$ and fundamental$_{\mathrm{CMB}}$. But then, the solution of the problem of communication via the super-particle view is blocked, for we arrive at two distinct spaces after all.

A possible solution to *this* problem, of course, would be to relax the requirements imposed on PO theories, such that either non-3D entities may figure as primitives, or that 3D entities are not required to be fundamental$_{\mathrm{ODEP}}$. However, construed as above, this is incompatible with the PO approach.

2.4.4 Disposition or Law?

So, taking the super-particle to be TBM's fundamental referent is of no avail to the problem of communication. We hence shall stick with the usual Bohmian particles. In this case, the second option for solving the problem of communication must be sought in the interpretation of the WF. In what follows, I will consider dispositional as well as nomological interpretations. So far, dispositionalism was assumed for TBM's actualizations of particle configurations, more or less implicitly, without pointing to an alternative reading. To make up for this, let me be explicit now: dispositionalism$_{TBM}$ is the view that, at any given time, the holistic N-particle system has a certain disposition, encoded in the WF, to actualize certain configurations. Note that the N-particle system is holistic in virtue of the particles sharing this disposition: they aren't completely independent of each other. Rather, the N-particle system exhibits a genuinely quantum-mechanical feature—entanglement—that doesn't supervene on the individual properties and spatial distribution of the individual particles. In other words: the N-particle system as a whole has a feature that can't be reduced to the features of the individual particles. The latter, thus, aren't ontologically independent of the former. On dispositionalism$_{TBM}$, the problem of communication is straightforwardly avoided: the 3N-dimensional WF encodes the disposition for the actualization behaviour of the N-particle system as a whole, represented as a point in 3N-dimensional configuration space. But because of that, dispositionalism$_{TBM}$ isn't compatible with the PO approach: the particles in 3-space aren't fundamental$_{ODEP}$, and the super-particle isn't 3-dimensional. Neither are suitable candidates for a primitive ontology.

When adopting dispositionalism$_{TBM}$, it seems, TBM isn't compatible with the PO approach. But dispositionalism certainly isn't the only option to interpret the WF. Chen (2019) provides a survey of several realist interpretations of the WF, amongst them nomological interpretations (ibid., 8): "nomological interpretations hold that the wave function is nomological, i.e., on a par with laws of nature." He distinguishes "two kinds of nomological interpretations of the wave function: the strong nomological interpretations and the weak nomological interpretations." Dürr et al. (2012, p. 12.3) discuss the possibility of the former for BM (ibid., 266): "one should think about […] the possibility that it's nomological, nomic – that it's really more in the nature of a law than a concrete physical reality." Likening it to equations of motion in classical mechanics, they suggest an analogous interpretation of the WF (ibid., 267), as "just a convenient device in terms of which the equations of motion can be nicely expressed. We're suggesting that you should regard the wave function in exactly the same way." Esfeld's "quantum Humeanism", following Chen, is an example of a weak nomological interpretation. Esfeld (2014, p. 455) holds that it is the primitive ontology approach to quantum physics that first of all enables quantum Humeanism: "the background that enables Humeanism to stand firm is the development of what is known as primitive ontology theories of quantum physics." Apart from the matter distributed in ordinary space(-time), there is, according to Esfeld, nothing but a law "as that what fixes (in a probabilistic or a deterministic manner) the temporal development of the distribution of matter in physical space, given an initial configuration of matter. That's all. In particular, the quantum mechanical wavefunction is part and parcel of the law instead of being a physical entity on a par with the primitive ontology." For his Humean approach, however, this law must not exist independently of the PO, but supervene on it (ibid., 456): "the Humean has to regard the law as supervening on the distribution of matter throughout the whole of space-time, that is, the entire mosaic of 'local beables' or local matters of particular fact." As such, it is part of the "best system" systematizing these "local matters of particular fact" (ibid., 459): "For the Humean, the universal wavefunction and the dynamical law that appear in a physical theory such as Bohmian mechanics or the GRW theory are part of the best system, that is, the system that achieves the best balance between being simple and being informative in capturing what there is in the physical world. That system – and everything that belongs to it – supervenes on the entire distribution of the local matters of particular fact throughout the whole of space-time."

It is neither my intent to arbitrate between different options of nomological interpretations of the WF, nor to advocate nomological interpretations in general. That being said, for those insisting on the PO approach, both, weak and strong nomological interpretations, are options worth considering for the interpretation of TBM's WF. Neither fall prey to the problem of communication, nor put non-3D entities into the primitive ontology. Imposing these demands, a nomological reading of the WF seems to be the most promising way to classify TBM as a PO theory (while solving the problem of communication). Is this a conclusive argument to the effect that TBM's WF must be read this way? Certainly not. For one, there might be other options beyond the ones presented in this paragraph. Second, the aim of the exercise was to show that TBM can be made compatible with the PO approach. It can, but that, in itself, is merely responding to a demand, rather than justifying it. Those who deem it crucial that a theory be compatible with the PO approach may go for a nomological interpretation. Others may adopt a different one, dispositionalism$_{TBM}$, for example, which provides a solution to the problem of communication by endowing the particles with a real, non-3D property, but, because of that, isn't compatible with the PO approach. I will abstain from arguing in favour or against the alternatives: in my view, the interpretational openness of TBM is not a bug, but a feature.

Irrespective of this interpretational issue, TBM's metaphysical adequacy can still be questioned. The problem we are going to discuss in the following section harkens back to the roots of TBM, Bell's "Everett (?) theory", and has been dubbed "temporal solipsism". Challenging the latter's metaphysical adequacy, it is the reason for Bell's dismissal of the theory, and hence deserves some attention. At first glance, it stands to reason that TBM, like Bell's Everett (?) theory, is undermined by temporal solipsism. That this is not the case, we will see now.

2.5 A Bit of History: Bell's Everett (?) Theory and Temporal Solipsism

Despite its empirical equivalence with BM, and its solution to the problem of communication, TBM still faces the objection brought forward by Bell (1987a, p. 136): "Everett's replacement of the past by memories is a radical solipsism – extending to the temporal dimension the replacement of everything outside my head by my impressions." This shall be called "temporal solipsism". Doesn't temporal solipsism undermine the adequacy of TBM? After all, the present configuration is not connected to any past or future configuration. So it seems that any talk of my existence past or future is vacuous, a mere illusion, just like, according to the ordinary solipsist, the external world is nothing but illusory. To address this issue, let's have a quick look at the historical origin of TBM—Bell's "Everett (?) theory"—and why Bell rejected it immediately.

The first outline of a theory similar to TBM is to be found in the last section (5) of Bell's *Quantum mechanics for cosmologists*. There, Bell (ibid., 133ff.) presents what he calls "the Everett (?) theory", which "will simply be the pilot-wave theory without trajectories".[20] He characterizes the Everett (?) theory as follows:

"Instantaneous classical configurations x are supposed to exist, and to be distributed in the comparison class of possible worlds with probability $|\psi|^2$. But no pairing of configurations at different times, as would be effected by the existence of trajectories, is supposed. And it is pointed out that no such continuity between present and past configurations is required by experience."

Here, Bell provides the general idea of a Bohmian theory without a GE. We are given the essential ingredients that make a theory Bohmian: possible configurations, i.e., configuration space ("the comparison class of possible worlds"), an instantaneous configuration x and the QEH for the distribution of ensembles of configurations as $|\psi|^2$. Despite not mentioning the GE—or rather, its omission—explicitly, he is eager to make clear that trajectories do not figure in the theory: configurations are taken to be instantaneous and "no pairing of configurations at different times, as would be effected by the existence of trajectories, is supposed."

After some remarks in which he compares his Everett (?) theory with the traditional account, Bell goes on to formulate a critique of the idea he just had introduced (136):

> In our interpretation of the Everett theory there is no association of the particular present with any particular past. And the essential claim is that this does not matter at all. For we have no access to the past. We have only our 'memories' and 'records'. But these memories and records are in fact *present* phenomena. [...] Everett's replacement of the past by memories is a radical solipsism – extending to the temporal dimension the replacement of everything outside my head by my impressions. [...] If such a theory were taken seriously it would hardly be possible to take anything else seriously.

After a quick comparison with creationism, Bell concludes his remarks on the Everett (?) theory. He quickly does away with, even ridicules the idea of a Bohmian theory without a GE (ibid.): "So much for the social implications. It is always

[20] To do actual history justice, it should be noted that, as Bell himself points out in a footnote, the origins of *Quantum mechanics for cosmologists* date back to an unpublished paper (*On the hypothesis that the Schroedinger equation is exact*, http://cds.cern.ch/record/956196) presented at Pennsylvania State University in September 1971, reproduced in Epistemological Letters in July 1978 and revised in Speakable and Unspeakable in Quantum Mechanics (1987a). Large parts of the final version were taken over from the original paper, including Bell's worries about temporal solipsism, although in the earlier version, the exact phrasing "radical solipsism" is not to be found.

interesting to find that solipsists and positivists, when they have children, have life insurance." It can hardly be argued that his treatment is thorough, taking his TBM-like idea seriously.[21] This is where Dürr and I disagree with Bell. We hold that TBM is not only empirically equivalent to QM—as discussed above and not contested by Bell—but that it is also metaphysically adequate. Temporal solipsism is not an issue. Showing the latter, i.e., showing that FAPP-continuous and consistent world histories can arise from a fundamentally stochastic, discontinuous micro dynamics, and explaining how this works, is the main objective of this book. In my presentation, I will expand on the presentation given in Dürr and Ehmann (2021). There, Dürr and I discuss Barbour's proposal of time capsules, replacing the latter with a coarse-grained notion of worlds in terms of Everettian world branches and supplement it with an argument about their quasi-persistence to tackle the issue of temporal solipsism. While I agree with the in-principle viability of the time-capsule approach, I will critically evaluate Barbour's construal of psychophysical parallelism, and the latter's compatibility with the claim that individual, instantaneous configurations are sufficient for the experience of sentient beings. I present a detailed account of how FAPP-continuous, consistent world histories emerge from a random, discontinuous micro dynamics that surpasses the presentation in Dürr and Ehmann (2021). But before I can do that, we need to have a look at Barbour's approach.

2.6 Barbour: Temporal Solipsism? Time Capsules!

Amongst contemporary philosophers of physics, Julian Barbour is probably the one who takes Bell's idea more seriously than anyone else. However, the timeless cosmology he develops in his 1999 book *The End of Time* departs significantly from both Bell's Everett (?) theory and TBM. In his theory, there is no universal time. Everything there is, is eternal and static. In particular, configurations aren't actualized sequentially, one "at a time", as is the case in the Everett (?) theory and in TBM. In Barbour's cosmology, there is no time, so configurations, too, are eternal and static. There is no configurational change over time (ibid, 39): "If we could see the universe as it is, we should see that it is static. Nothing moves, nothing changes". Accordingly, experience of time has to arise from within such eternal, static configurations (ibid., 53): "The instant is not in time, time is in the instant." To account for the experience of time flow, of things moving, he therefore introduces the notion of time capsules (ibid.): "I shall outline, through the notion of time capsules, a

[21] Barbour (1999, p. 300f.) agrees with this judgment about Bell's premature dismissal of the theory: "This is all very entertaining—and I too have children and life insurance—but these are just the kind of *ad hominem* quips that were tossed at Copernicus and Galileo. I do believe that Bell came close to a viable cosmological interpretation of quantum mechanics, and should have kept faith with his title ('Quantum mechanics for cosmologists'). But he left the cosmologists with nothing."

theory of how a static universe can nevertheless appear to teem with motion and change."

Neither Dürr and I in our paper on TBM, nor I in this book, commit to Barbour's assumption of a fundamentally static universe, in the sense of a universe without fundamental, universal time. For TBM, we assume universal time. At each point in time, one configuration is actualized. As opposed to Barbour's cosmology, TBM's configurations are not eternal. Despite this important difference, Barbour's time-capsule approach is a candidate for tackling the issue of temporal solipsism. Barbour is well aware of Bell's Everett (?) theory and his criticism regarding temporal solipsism (cf. ibid., 300). In fact, he devotes an entire section of his book (299ff.) to the theory. Therein (299), he identifies Bell's notion of instantaneous configurations containing records with his notion of time capsules: "[...] configuration points that can be called records. Although Bell did not use my term, such points are manifestly time capsules." More generally, he introduces time capsules as follows (31): "Any static configuration that appears to contain mutually consistent records of processes that took place in a past in accordance with certain laws may be called a time capsule."

One must think of such time capsules as configurations containing within them the complete history of a process up until that configuration, in the form of "snapshots". Barbour (ibid., 292) provides the following analogy: "Doting parents take daily snapshots of their child and stick them day by day into a progress book. The progress book after each successive day is like each successive point along the track in the big configuration space: it is the complete history of the child up to that date." Other examples of records present in time capsules are geological formations (30)—"all geological formations, rock strata in particular, are now invariably interpreted by geologists as constituting a record (to be interpreted) of past geological processes"—or the particle tracks in a cloud chamber (299)—"alpha-particle tracks [...] have the obvious interpretation that they are records of alpha-particle motion"—all telling of a coherent history of events, apparently following some principles of time evolution that can be described in terms of laws of nature.

In addition to their structural richness and internal consistency, time capsules are supposed to be "sought out" by the WF (300): "[Bell's] 'Everettian' interpretation is this: time exists, and the universal wave function Ψ evolves in it without ever collapsing. Because Ψ has the propensity to seek out[22] time capsules, it will generally be concentrated on them. Real events are actualized as follows. At each instant of time, Ψ associates a definite probability [...] with each configuration. At any instant, just one event is actualized at random in accordance with its relative

[22] Barbour's talk of the WF having "the propensity to seek out time capsules" is a bit obfuscating. (He repeatedly uses this metaphor, cf. Barbour, 1999, pp. 251, 296, 298, 300, 308, 316, 324.) What he means is that the probability density associated with the universal WF is concentrated over configurations that are time capsules, such that, at any time, the probability to be actualized is much higher for configurations that are time capsules than it is for configurations that are not time capsules.

probability. The higher the probability, the greater the chance of actualization. Since time capsules have the highest probabilities, they will generally be selected."

2.6.1 *Explaining Experiences*

For his own, timeless cosmology, Barbour's demand that time capsules are predominant is intertwined with "an assumption in the hallowed tradition of Boltzmann: *only the probable is experienced*." (ibid. 266, his emphasis.) This assumption is controversial. When is a configuration probable enough to be experienced? Why, above all, should it be the case that only highly probable configurations are experienced at all? Barbour provides no good justification for his assumption. Nevertheless, at least at first sight, he has reason to make it despite: the vast majority of configurations in configuration space doesn't admit of any experiences at all, and of those that do, the vast majority doesn't admit of the reconstruction of FAPP-classical, coherent histories. Given that all configurations in his timeless cosmology are eternal, they are all actualized "at the same time" (which, in a timeless cosmology, can only be a figure of speech), which means that all possible experiences are actual. But if all possible experiences are actual, then by far most actual experiences are of incoherent histories. Yet, obviously, our own experiences are of the ordered kind—of consistent memories and records, admitting of the reconstruction of FAPP-classical, coherent histories. This seems unexpected. Our experiences, by all likelihood, should be incoherent.

Barbour's assumption that the coherent time capsules are predominant, and that only the predominant configurations are experienced, allows him to reconcile the expectations derived from the theory with our experiences. Prima facie, this explains why we have the experiences we happen to have.

Given one demands this kind of explanation from Barbour's timeless cosmology, one should also demand it from TBM. Other than Barbour, however, in order to provide such an explanation, we don't have to assume that only the probable is experienced. On TBM, only one configuration is actual at a time. No matter how improbable its actualization is, once it is actualized, and contains experiencing beings, then these beings will have the experiences they just happen to have.[23] So when we demand for TBM that time capsules are assigned the highest probability to be actualized by the WF, what we want to achieve is that those configurations admitting of coherent histories are predominantly *actualized*, as opposed to being the only ones admitting of experiences in the first place. On TBM's sequential actualization dynamics, this is enough to reconcile theoretically derived expectations and our actual experiences. Barbour's controversial assumption isn't required.

---

[23] This is true under Barbour's further assumption that individual, internally static and eternal time capsules are sufficient for experiences to arise. A critique of this assumption will be spelled out later. For the present purposes, we can safely assume that it holds.

Except for this difference, the rest of the reasoning goes quite similar for both theories. Time capsules should be assigned the highest probability to be actualized by the WF, because otherwise, e.g., assuming an equidistribution of probability over the totality of configuration space, almost all actualized configurations containing records at all would be configurations containing records that don't allow for the reconstruction of coherent histories, for such configurations vastly outnumber the rather special time capsules. As Barbour puts it (ibid., 289): "among all possible worlds, the dull, disordered, incoherent states are overwhelmingly preponderant, while the ordered states form a miniscule fraction. But such states, sheer implausibility, must be presupposed if history is to be made manifest".[24] Of course, by far the largest portion of configuration space consists of configurations that don't contain any records from which histories, coherent or not, could be reconstructed. The primary reason for demanding that time capsules are assigned the highest probability is not to ensure that in most actualizations, records are present, but to ensure that most actualizations in which records are present admit of a coherent history.

We must also revisit the issue of psychophysical parallelism: given strict dualism is not assumed, there has to be some "physical counterpart of psychological experience" (Barbour, 1999, p. 302). Barbour follows Bell in his assumption that this is to be found in the configuration, i.e., in particle positions. From this assumption, he draws the following conclusion (ibid., 302, my emphasis): "it is clear from the way [Bell] makes memories and records responsible for our idea of the past, rejecting any 'thread' connecting configurations at different times, that subjective awareness of both positions and motions of objects must be derived from the structure in *one instantaneous* configuration. The self-sentient configurations must be time capsules. Not only the kingfisher but also the appearance of its flight must be *in one configuration*, for nothing else would be logically consistent". Later on, I will contest this. It is not the case that subjective awareness of motions must arise from *one* configuration, and that nothing else would be logically consistent. I will argue that Barbour's claim contradicts our best understanding of the emergence of mental

[24]The situation is reminiscent of the problem of Boltzmann brains. Given that the present state of our universe results from a random fluctuation out of equilibrium, we should expect ourselves to be Boltzmann brains: beings endowed with the necessary structure to hallucinate our experiences—including records, memories and anticipations of the future—but with not much more, for fluctuations into states that only contain such minimal beings hallucinating ordered structure and coherent records are much more likely than fluctuations into states where all the ordered structure and coherent records we seem to observe actually exist. (Cf. Carroll, 2017.) Even worse: by far most Boltzmann brains fluctuating into existence should be "batty", such that the structures and records they hallucinate are disordered and incoherent. (Cf. Norton, 2015.) Likewise, if TBM's actualization probability was equal for all configurations, we should expect most sentient beings to be Boltzmann brains, and most Boltzmann brains to be of the batty type. In fact, the analogy is quite to the point. Just replace the random fluctuations as supposed for the Boltzmann-brain scenario with random actualizations as featured in TBM. Given equal probability, of those randomly actualized configurations only few contain beings capable of experiences. And of those, in turn, almost all contain only batty Boltzmann brains. Were it not for the probability density to be distributed nonuniformly, we indeed should expect that amongst all brains instantiated, almost all are batty Boltzmann brains. Again, this is not logically ruled out by our experiences, but unexpected.

phenomena, and that he is aware of that: for all we know, the sentience of a sentient being arises not from one of its configurations, but from processes, involving change between several configurations. Barbour, pointing to the relevancy of brain processes (cf. ibid., 266), expresses a similar view. While he is right that upon "rejecting any 'thread' connecting configurations at different times, [...] subjective awareness of both positions and motions of objects must be derived from the structure in *one instantaneous* configuration", I reject the antecedent for TBM: even assuming a stochastic, discontinuous micro dynamics as features in TBM, on a higher-level description, a "thread connecting configurations" can be had, namely in the form of world histories.

But first things first. For this introduction it shall suffice to assume that the criterion of psychophysical parallelism is met, i.e., that sentient beings arise from time capsules in such a way that a singular, instantaneous configuration is sufficient for their subjective experience. Given the probability density indeed concentrates on time capsules, we can expect that the experiences and memories of most actualized beings are of the regular, consistent kind, like ours are. Our history in its regularity, then, is nothing special.

But how is this predominant actualization of time capsules spelled out in TBM? Here comes into play the branching of the WF into FAPP-classical world branches, induced by decoherence, which accounts for the many-worlds character TBM inherits from the Everett interpretation.[25] With the WF being concentrated on FAPP-classical world branches, the actualization probability of configurations in the support of FAPP-classical world branches is overwhelmingly large compared to those in the support of non-classical branches. It is those configurations in the support of FAPP-classical world branches that contain mutually consistent records admitting of coherent histories, and that are called time capsules in Barbour's terminology. Now, since the actualization probability of configurations in TBM is given by the probability density $|\Psi|^2$, the super-particle, over the course of many actualizations, traces out FAPP-classical world branches, with overwhelmingly high probability. In other words, it almost always instantiates time capsules. We mustn't wonder why we find ourselves in such a "special", structurally rich configuration with consistent records, for, although such configurations might be special in the sense that they take up only a tiny volume in configuration space, their *actualization*, taking into account the branching structure of the WF and the resulting actualization probabilities, isn't special at all.

[25] Everett's heritage is the many-worlds feature, not decoherence per se. The latter is by no means exclusive to Everett. Cf. Schlosshauer (2007, Ch. 8), discussing the role of decoherence for several (types of) "interpretations" (read: interpretations and alternative theories), or Bacciagaluppi (2016).

2.6.2 *Fine-Tuning for Life and Rare Time Capsules: An Analogy*

The arguments presented above made use of the probability density being concentrated on configurations admitting of coherent histories in order to explain the regularity of our experiences. However, it can be put into question whether such explanations are necessary and called for.

Let me try to flesh this out with an analogy between the rarity of time capsules and the fine-tuning for life of our universe. The parameters (e.g. constants of nature) defining our universe seem to be fine-tuned for life. Friederich (Friederich, 2018, 1.1.1) provides, amongst others, the following example of fine-tuned constants of nature: "The strength of gravity, when measured against the strength of electromagnetism, seems fine-tuned for life [...] If gravity had been absent or substantially weaker, galaxies, stars and planets would not have formed in the first place. Had it been only slightly weaker (and/or electromagnetism slightly stronger), main sequence stars such as the sun would have been significantly colder and would not explode in supernovae, which are the main source of many heavier elements [...]. If, in contrast, gravity had been slightly stronger, stars would have formed from smaller amounts of material, which would have meant that, inasmuch as still stable, they would have been much smaller and more short-lived".

Now consider the space of *possible* universes—a parameter space in which each relevant parameter is represented by a dimension. For example, the gravitational constant is represented by a dimension, and can vary over a range of different values.[26] Likewise other universal constants. Each point in this parameter space represents a possible universe.

This parameter space, then, comprises a huge "number" of possible universes, all set up with different values for the relevant parameters. That only "few" (a small volume in parameter space, assuming there is an appropriate measure for that) of these possible universes admit of life like ours is the essence of the claim that our universe is fine-tuned (cf. Barnes, 2012, p. 529). For our purposes of constructing an analogy, we may assume that this is indeed the case, as suggested by Friederich's example. Under these conditions, if one actual universe would be selected at random from the vast space of possible universes, it would be highly improbable that this randomly selected universe is one of the few capable of harbouring life as we know it. This, as Barnes (ibid.) puts it, is what makes the claim that our universe is fine-tuned for life interesting: "it makes the existence of life in this universe appear to be something remarkable, something in need of explanation. The intuition here is that, if ours were the only universe, and if the causes that established the physics[27]

[26] The terminology sounds a bit oxymoronic. How to imagine a *varying constant*? Of course, in this context, being a varying constant means being (effectively) constant in a given universe, but varying over parameter space.

[27] Barnes calls the parameters specifying a universe its "physics".

of our universe were indifferent to whether it would evolve life, then the chances of hitting upon a life-permitting universe are very small."

Standard explanations of fine-tuning for life are the deliberate creation of our universe by an intelligent designer, or the actual existence of a multitude of different universes. I am not going to engage in the debate about which to prefer. The point here is to make an analogy: on the one hand, a space of possible universes, only few of which admit of life; on the other, a space of possible configurations, only few of which admit of coherent histories. For TBM, we assume that this latter space is actually explored; the super-particle jumps through configuration space. Accordingly, to construct the analogy, we assume that the former space is explored, too; a multitude of different universes from the space of possible universes is brought into being.[28]

Given this multitude of universes is actual, it isn't surprising that the universe we inhabit, featuring just the right set of parameter values, exists. In the space of possible universes, there are universes featuring different sets of parameter values, most of them not allowing for any sentient beings at all. We certainly inhabit a special universe, with parameters suitable to us. It indeed is fine-tuned for our existence, but it is just one amongst the many possibilities explored. Further explanation for its fine-tuning seems called for only if we assume that there is only one universe. But if we assume many universes, as we do for this analogy, it is to be expected that some actual universe(s) will have the right conditions to bring about life like ours. Analogously, if we assume many configurations being actualized, as we do for TBM, no special concentration of the probability density over time capsules is necessary in order to explain our experiences. As it turns out, the arguments above, involving a special concentration of the probability density, attempt an explanation for something that is already explained.

Let's pause here for a moment. I have just pushed back against the idea that an explanation of our experiences must be given in terms of a special concentration of the probability density. But, in a sense, I still gave an explanation, albeit a different one, drawing on the fact that, on TBM, many configurations are actualized, one after another. One might wonder whether *any* explanation of us having the experiences we happen to have is needed *at all.*

2.6.3 Explaining the Need for Explaining Experiences: A Conjecture

Certainly, there isn't always a need to explain the occurrence of an unlikely event, like the actualization of the time capsule we inhabit, or the existence of the fine-tuned-for-life universe we inhabit (Friederich, 2018, 2.2): "Even if fine-tuned

[28] As a side note, for the following argument, it doesn't matter whether these universes exist in parallel or one after the other.

conditions are improbable in some substantive sense, it might be wisest to regard them as primitive coincidences which we have to accept without resorting to such speculative responses as divine design or a multiverse. It is indeed uncontroversial that being improbable does not by itself automatically amount to requiring a theoretical response. For example, any specific sequence of outcomes in a long series of coin tosses has low initial probability (namely, 2^{-N} if the coin is fair, which approaches zero as the number N of tosses increases), but one would not reasonably regard any specific sequence of outcomes as calling for some theoretical response".

Indeed, being surprised by the occurrence of an improbable event and then inferring that this event likely is one of a sufficiently long series of past events would be falling prey to the inverse gambler's fallacy, originally described by Hacking (1987). Rowbottom (2015, IX.1) characterises it as follows: "[The inverse gambler's fallacy] involves mistakenly inferring what happened in the *past* [...] because of what happens in the present. Or more precisely, in game-like scenarios, it involves the assumption that results at some point in time are *dependent* on earlier results in some way." In other words, the (false) premise at work when committing the inverse gambler's fallacy is.

(IGF*) If an unlikely event occurs, then it likely is part of a long series of past events.

Given that an unlikely event occurs, and IGF* is assumed, the (false) conclusion

(C*) The unlikely event likely is part of a long series of past events follows. In light of this, it seems best to abstain from any attempts at explaining unlikely events by inference to a past series of events.

That being said, the occurrence of unlikely events, at least psychologically, can be surprising, and people tend to feel the need for an explanation.[29] Let me try to make sense of this apparent need for explanation, for the case of our universe's fine-tuning for life, and, by analogy, for our time capsules "fine-tuning" for the experience of a coherent history, exploiting Friederich's coin-tossing example.

In a series of ten tosses of an ideal, fair coin, I should not expect any *specific* sequence over any other *specific* sequence. The specific sequence HHHHHHHHHH, or the specific sequence HTHTHTHTHT, is just as likely as the specific sequence THTTHHHTTH (which I have just generated tossing an actual coin). For each individual coin toss, H and T both have probability $\frac{1}{2}$, independently of the other tosses, so each specific sequence of ten coin tosses has probability $\left(\frac{1}{2}\right)^{10} = \frac{1}{1024}$. They are all equally (un)likely. I'd commit the inverse gambler's fallacy if I inferred from a sequence of all heads that it likely is part of a long series of sequences.

However, flipping a coin ten times, as I did, I'd still be quite surprised by the outcome HHHHHHHHHH, while I wasn't surprised by THTTHHHTTH, the actual result I got. The former sequence triggers the (psychological) need for an

[29] There is some empirical evidence suggesting that in fact, people tend to commit the inverse gambler's fallacy, see Oppenheimer and Monin (2009).

explanation. Why is that so? Both specific sequences have the same probability. I really shouldn't be more surprised by one than by the other. One really shouldn't require an explanation while the other doesn't.

Here is my conjecture. The reason that I am surprised by HHHHHHHHHH but not by THTTHHHTTH might have to do with coarse-graining sample space into sets, resulting in a partition of sample space, in this case, into equivalence classes. Let me explain. Glossing over the specific order of Hs and Ts in sequences, i.e., only being concerned about the relative frequency of Hs and Ts, the latter sequence above is representative of a much larger set of sequences than the former. I'm surprised not because I got one specific sequence instead of another, equally likely sequence, but because I got a sequence from a tiny set instead of a sequence from a comparatively large set. (With all possible sequences being equally likely, the probability of ending up with a sequence from a certain set is the probability of ending up with a specific sequence times the number of sequences in that set.) There are 252 sequences containing exactly five Hs. They form an equivalence class. No other equivalence class (with the equivalence relation "containing the same number of Hs") in the sample space of possible sequences is larger. So, having five Hs in ten tosses is the expected frequency. There are 210 sequences with exactly four Hs, and 210 with exactly six. 120 each with three or seven Hs, and so on. What's so surprising about the occurrence of the all-heads sequence, I figure, is that there is only one sequence in its equivalence class, while other equivalence classes are much bigger. The further the relative frequency of H's and T's deviates from the expected frequency, i.e., the smaller the relative size of the equivalence class the sequence is part of, the more surprised I am. My surprise is rooted not in the fact that a *specific* sequence occurred instead of another *specific* sequence, but in the fact that a sequence from the smallest equivalence class occurred instead of one from the bigger classes.[30]

I can imagine that this is why examples in presentations of the inverse gambler's fallacy usually are set up as they are. Hacking, e.g., construes an example with a gambler inferring from a roll of two dice both showing sixes that this roll most likely was part of a longer series of rolls. Why does Hacking construe the example with double sixes? Why not with the first die showing a two and the second showing a four, say? It's just as likely a combination as two sixes are. However, an example with such a seemingly common combination would feel strange, wouldn't it? Why does it seem common, while double sixes don't? My hunch is that we're coarse-graining the sample space of possible outcomes into equivalence classes with respect to "being a double". Then all the double somethings are put into a set, and all the rest into a different set, and since the former set is much smaller, the double

[30] Even sequences with the expected frequency might psychologically be surprising. For example, the sequence HTHTHTHTHT, or HHHHHTTTTT. These look surprisingly ordered, as if they had been generated by applying a (deterministic) rule. So, to provide a tenable account for why some outcomes are psychologically surprising while others aren't is probably a bit more complicated than my simplified example suggests. However, for conveying the basic idea and making my point, it should be enough.

somethings look surprising and in need of an explanation, motivating the unwarranted inference.

Let's look at another, even more illustrative example. Suppose a dart is randomly thrown once at a board with 32 × 32 equally sized squares. Only one of them is red, all other 1023 squares are green. Each square has the same probability of being hit as any other. But glossing over the fact that the squares are individual squares and partitioning the board under the equivalence relation "having the same colour" results in two equivalence classes, one much larger than the other. Suppose that the dart landed on the one red square. It isn't surprising that it landed on any *specific* of those 1024 squares. (If the squares all had different colours, or all the same colour, none would stand out.) But it is somewhat surprising—at least I would be surprised—that it landed on the single red square, and not on *any* of the 1023 green ones. This is what reeks of conspiracy.

Now, the question seems to be whether my surprise, on which my suspicion of conspiracy and demand for an explanation are based, is justified or not. Am I justifiably surprised that the dart landed on the red square? This, it seems, is a matter of perspective. Looking at the squares individually, on the "lower", not coarse-grained level of description, the situation presents itself as an equidistribution; all squares have the same probability of being hit. Because of that, being surprised by a specific square being hit isn't justified. On the higher, coarse-grained level of description, i.e., on the level of colours, probabilities aren't equidistributed; not all colours have the same probability of being hit. But if being surprised by an unlikely outcome, given there is a much more probable alternative (as opposed to many equally unlikely alternatives), is justified in general, then being surprised by the red square being hit, given all other squares are green, is also justified.

This puts me in an almost paradoxical situation. One and the same event, a specific square being hit, seems surprising or unsurprising, depending on whether I choose to coarse-grain for colour or not, i.e., on the partition I impose on sample space. If I do coarse-grain for colour, then I expect a green square to be hit almost always. The red square being hit on a single throw thus triggers a need for explanation.[31] If I do not coarse-grain for colour, I don't expect any outcome over any other outcome, and thus won't ask for an explanation. But with my expectations depending on whether I am coarse-graining or not, prospects of justifying my surprise seem forlorn. That is, unless the partition I impose itself is justified—if it is a "natural" partition, so to speak.

[31] A possible explanation for why the red square was hit, and not any of the green ones, is that the dart wasn't thrown randomly after all. It was Peter 'Snakebite' Wright, the 2022 World Darts Champion, deliberately aiming for the red square. Another possible explanation is that the dart in fact was thrown randomly, but many times over. Throwing often enough, it will eventually land on the red square. In both these cases, I wouldn't be surprised by the red square being hit. Interestingly, both explanations involve me being mistaken about the initial setup: to make them work, I must revoke my assumption that the dart is thrown at random, or that it is thrown only once, respectively, and replace it with a different assumption. Note that assuming a proposition isn't the same as inferring it. In particular, assuming that the dart was thrown many times isn't to be mistaken for inferring that it was thrown many times.

To get terminology out of the way, I call a partition "natural" iff it recovers actual, higher-level similarities and differences between the elements of sample space. To be more precise, a partition is natural iff it groups together those and only those elements that are equivalent with respect to an actual, higher-level property. For example, the partition of the dartboard into colour equivalence classes is called natural because it recovers the actual, higher-level similarities and differences in colour of the squares: all and only red squares are grouped together in a set, all and only green squares are grouped together in another set.

You can see where this is going. Arguably, coarse-graining the parameter space of possible universes with respect to hospitability for life, or coarse-graining TBM's configuration space with respect to suitability to bring about the experience of a coherent history, result in natural partitions, too. Like the colour partition of the dartboard, they recover actual, higher-level similarities and differences between the elements of these spaces. Like with the dartboard, I shouldn't be surprised that any one of all the possible universes from parameter space was selected over any other. But being surprised that the one and only actual universe is hospitable to life seems to be justified.

So much for my conjecture about why certain outcomes seem surprising and psychologically demand for an explanation, while, when looking at the underlying sample space, all outcomes are equally likely. Even assuming the actualization probability density being flat over TBM's configuration space, the fact that my configuration was actualized does not demand for an explanation any more than the actualization of any other possible configuration. That I am still surprised by this actualization, and thus demand an explanation for it, is due to the partition I impose on configuration space: there are possible configurations, called time capsules, that allow for the experience of a coherent history, and there are other possible configurations that don't. Coarse-graining with respect to suitability to bring about the experience of a coherent history recovers these higher-level similarities and differences. The resulting partition is thus natural. It tells me that only a small set of possible configurations are time capsules, while all others aren't, and that time capsules thus are much less likely to be actualized than non-time capsules. Assuming there is only one configuration actualized, I'd feel very justified in being surprised that this one configuration was a time capsule, despite the fact that I know that each specific configuration is as likely to be actualized as any other (assuming actualization probability density being flat over configuration space).

Being justifiably surprised of, and thus demanding an explanation for, the actualization of a time capsule at some specific point in time, may I justifiably infer that, most likely, many configurations have been actualized before it, i.e., that the actualization of that time capsule most likely is part of a long series of actualizations, or would I fall prey to the inverse gambler's fallacy? Indeed, I would. Said time capsule could be the first configuration ever actualized, or the fifth, or the 10^{679} th. Luckily, however, I don't need to make this inference. The dynamics of TBM already tells me that uncountably many configurations are actualized in any finite time interval. This is built into the theory. I don't need to infer the likely existence of a series of previous actualizations, so the inverse gambler's fallacy is straightforwardly avoided.

2.6.4 Is TBM Empirically Incoherent?

On TBM's dynamics, the occasional actualization of time capsules is expected, even without a special probability density distribution. However, the decoherence-induced branching structure of the WF is relevant in a different matter. Barrett (1996) brings forward the charge of empirical incoherence against Bell's Everett (?) theory. He calls a theory empirically incoherent (ibid., 50) if it "fails to predict the existence of reliable records of an observer's measurement results to which the observer has epistemic access", and claims that the records present in time capsules are indeed unreliable, in the sense that the records contained within a given time capsule, e.g., our present one, typically are incompatible with records in previously actualized time capsules (ibid., 55): "the current particle configuration is independent of any past or future configurations [...] This means that one's records of measurement results would typically change in a pathological way over time and hence be wildly unreliable as records of what actually happened." He provides the following example:

"Suppose [...] that an observer M measures the x-spin of a spin-1/2 system S that is in a superposition of x-spins [...]. Suppose that the observer gets the result x-spin up for the outcome of her first measurement. Now what happens if the observer carefully repeats her measurement? The linear dynamics tells us that the quantum-mechanical state after the second measurement will be

$$\alpha|\uparrow,\uparrow\rangle_M|\uparrow\rangle_S+\beta|\downarrow,\downarrow\rangle_M|\downarrow\rangle_S$$

So what is the classical configuration after this measurement? [...] If M does in fact get 'x-spin down' for her second measurement, the classical configuration will now be one associated with the second term of the above state, which means that M's 'record' of her first measurement will now read 'x-spin down', and it will thus appear, based on an examination of her records, that her two measurements did in fact yield the same result. More generally, one can show that the classical configuration would almost always be such that one's records would exhibit the statistical correlations predicted by the standard theory whenever it makes unambiguous predictions."

Barrett regards Bell's Everett (?) theory—and a fortiori TBM—as empirically incoherent, because, at least prima facie, it's records can be expected to be unreliable in this sense. However, he seems to overlook the import decoherence has on this matter. With decoherence, the WF branches into FAPP-classical world branches. Configurations are actualized at an arbitrarily high rate (see below), so consecutive actualizations of different time capsules within a world branch, because of the latter's FAPP-classicality, can be expected to be mutually consistent with respect to the records they contain. So, contrary to Barrett's fear, at least within a given branch, TBM's records are epistemically accessible and reliable. They won't be illusory in the sense of providing evidence contradicting what actually happened in the past. TBM thus satisfies Barrett's requirements for empirical coherence. The charge of empirical incoherence is avoided.

2.7 Temporal Solipsism Solved?

This brings us back to the issue of temporal solipsism. Recall Bell's complaint that "if such a theory were taken seriously it would hardly be possible to take anything else seriously". One might still think that the theory is metaphysically deficient. After all, considering the vastness of configuration space, it is very unlikely that *precisely* the one time capsule I inhabit right now, this unique configuration, will be actualized again, even if the probability density is concentrated on time capsules, and consecutive time capsules within a world branch contain mutually compatible records. Consequently, I should exist for only one instant, such that all my memories of the past and anticipations of the future are rendered mere illusions. Even worse, not only my memories, but all records are meaningless—bedeutungslos in Frege's sense[32]: they point to a past that never existed. Annual rings in trees, the pictures of me as a child, geological formations including the fossils found in them, they all have been created at one instant, and cease to exist immediately after. In light of this, Bell likening the Everett (?) theory to creationism seems understandable.

However, this problem arises only if time capsules are identified with individual, unique configurations, such that for any pair of actualized configurations $Q_{i,j}$, with $Q_i \neq Q_j$, different time capsules $TC_i \neq TC_j$ are instantiated, which is the case according to Barbour's presentation of Bell's Everett (?) theory (302): a time capsule is supposed to be "one instantaneous configuration". By contradistinction, "worlds", as mentioned earlier, are a coarse-grained concept. To provide a preliminary delineation: unlike time capsules, worlds do not consist of only one configuration. Instead, worlds are regions of finite volume in configuration space, comprised of all configurations instantiating macro states that are indistinguishable from a given macro state, i.e., worlds result from coarse-graining with respect to macroscopic features. The procedure will be explained in greater depth later.

Let's denote such a macro world $\mathcal{W}$, and the fact that one of its configurations is actualized at some time t by $(Q \in \mathcal{W}, t)$.[33] Since the set of configurations in $\mathcal{W}$ is a finite region in configuration space, the probability of some configuration from

[32] Cf. Frege (1892).

[33] t, here, refers to universal time, i.e., the time parameter of the SE, describing the evolution of the universal WF.

$\mathcal{W}$ being actualized at a specific $t = t_i$ depends on the size of that region and the probability density as given by the BR for that region and time:

$$P(Q \in \mathcal{W}, t_i) = \int_{\mathcal{W}} d^{3N}Q \left|\Psi_{t_i}(Q)\right|^2$$

Because worlds correspond to a region of *finite* volume in configuration space, this probability might be small, but is always non-zero (given the probability density over this volume isn't zero):

$$\forall t : 0 < P(Q \in \mathcal{W}, t)$$

Now take some finite time interval $I \subset \mathbb{R}$. With the probability of $Q \in \mathcal{W}$ being actualized at any arbitrary time t already being greater than zero, the probability of it being actualized within the time interval I, consisting of uncountably many instants, must be greater than zero as well: $P(Q \in \mathcal{W}, I) > 0$.

We can then select at random some T instants from I: $t_1, \ldots, t_T \in I$. Think of T as the number of instants at which some configuration is actualized within the given time interval I, such that $R = \frac{T}{I}$ is the actualization rate (in actualizations per time span) for the time interval I. For example, if we select $T = 1000$ instants from a time interval I of 1 s, the actualization rate is $R = \frac{1000}{s}$. Note that $R \to \infty$ for $T \to \infty$ and finite time intervals. Because actualizations of configurations in general are stochastically independent—recall that $P(Q_t \cap Q_{t'}) = P(Q_t) * P(Q_{t'})$—actualizations of configurations within a given world at different times are stochastically independent as well: $P(Q \in \mathcal{W}, t \cap Q \in \mathcal{W}, t') = P(Q \in \mathcal{W}, t) * P(Q \in \mathcal{W}, t')$. Despite this stochastic independence, we can expect that configurations from our world are actualized time and time again, contra Bell's main worry: with the actualization probability $P(Q \in \mathcal{W}, t_i)$ for a configuration in $\mathcal{W}$ to be actualized at t_i, the expectation that some $Q \in \mathcal{W}$ is actualized during I is the sum over the individual actualization probabilities at each instant t_i of the selected T instants $t_1, \ldots, t_T \in I$:

$$E[Q \in \mathcal{W}; t_1, \ldots, t_T] = \sum_i P(Q \in \mathcal{W}, t_i) = P(Q \in \mathcal{W}, t_1) + \ldots + P(Q \in \mathcal{W}, t_T)$$

As can be seen, this expectation depends on the number T of actualizations during I, i.e., on the actualization rate. If $R \to \infty$, $E \to \infty$ as well:

$$E[Q \in \mathcal{W}; t_1, \ldots, t_T] \overset{R \to \infty}{\to} \infty$$

So, under the assumption that, during a finite time interval, arbitrarily many configurations are actualized, configurations from our world $\mathcal{W}$ can be expected to be actualized arbitrarily many times in the same, finite time interval as well.

What's the import of these considerations on our question at hand? Is Bell's worry regarding the alleged temporal solipsism of TBM off the table? Indeed, as we have seen, on TBM's construal of worlds, configurations within a world aren't actualized just once. In particular, configurations from the world inhabited by you and me are actualized infinitely many times. We can be reassured that they will be actualized again, and already have been actualized before, many times over. So, this particular worry is averted.

However, temporal solipsism can be understood as comprising more than the fear that any particular configuration is instantiated only once. Concerns may also arise about the temporal connection between different configurations at different instants of time. While Bell's original unease was particularly focussed on the "replacement of the past by memories" (Bell, 1987a, p. 136), he at least hinted at a similar issue regarding the future, recall his comment about solipsists having life insurance (ibid.). His worries about the past being a mere illusion carry over to future configurations as well: in both directions of time, the connection between different configurations, in particular temporally *remote* ones, seems problematic. Recall that time capsules in Barbour's account are supposed to be static (Barbour, 1999, p. 31): "Any static configuration that appears to contain mutually consistent records of processes that took place in a past in accordance with certain laws may be called a time capsule." But Barbour, by his claim that external, universal time does not exist, not only is forced to speak of *static* time capsules in the sense of *internally unchanging* configurations. Individual configurations are static in this sense by construal. Other than that, configurations are also supposed to be *static* in the sense of being "*not in time*" or "*eternal*". So there are no time capsules at *different times* in Barbour's account. Yet, assuming for TBM, by contrast to Barbour's timeless cosmology, that there are different configurations actualized at different times, there are temporally remote worlds, for example the world instantiated right now and the temporally remote world where I have children and life insurance, and the even more remote world where my future family benefits from said insurance. Likewise, the past world where I, as a young boy, am sitting in my bedroom playing with Lego and video games, knowing nothing about Barbour's concept of time capsules or even his book, which hasn't been written at the time. The world I inhabit now contains records of this earlier one—memories, photographs, tape recordings. However, the relation is not symmetric. As Butterfield (2002, p. 319) puts it: "one time capsule can contain records of another, without the other similarly containing records of the first". There was no record of my future self contained in the old world, just as there are no records of my children in the world I inhabit now. How are these instantiations of different worlds at different times linked up, over significant time spans in particular?[34] On their own, neither the fact that configurations

[34] Sticking with a timeless view, the question would have to be rephrased: How are worlds at different positions in configuration space linked up, such that they form a common history? For example, how is the world with my tea cup sitting on my desk related to the world with my tea cup on the floor, shattered into pieces? The answer Barbour would give is that the world with the broken cup contains records of the world with the unbroken cup—and likewise, for explaining how a

from our world are actualized many times over, nor that consecutive time capsules within a world branch contain mutually consistent records, provides an answer to this question. In our paper, Dürr and I have tackled this issue by abandoning the original, Barbourian notion of static time capsules and resorting to Everettian world branches, which, for different instants of time, contain different, instantaneous configurations in their support, but themselves must not be considered instantaneous or static: they reach over extended time intervals, and hence comprise a history of configurations. And since world branches are a consequence of decoherence,[35] this history is FAPP-classical, i.e., internally consistent. Because of that, records of the past, and likewise anticipations of the future, do point towards real instantiations of our world branch (Dürr & Ehmann, 2021, p. 180): "They permit inferences to our macro-world's history; within each atto-second, infinitely many temporal fragments of it are realised. On TBM, historical records aren't illusory or deceptive: they correctly describe a branch of reality (quite literally: an Everettian branch – that is, a quasi-classical history)." With world branches not only being instantiated many times over, but also forming quasi-classical histories, Bell's temporal solipsism is warded off. While the fundamental picture still comprises independent, discontinuous actualizations of configurations, on a higher level of description, world branches (quasi-)persist. Figure 2.1 provides an illustration of the quasi-persistence of world branches according to TBM.

This approach needs to be spelled out in greater detail than could be done in Dürr's and my paper on TBM. Combining elements of Barbour's time-capsule approach with elements from the Everettian theory, worlds correspond to finite volumes in configuration space. They are sets of (possible) configurations all of which (counterfactually) instantiate a macro state that is indistinguishable from a given macro state. Worlds, in some respect, resemble Barbourian time capsules, yet are importantly different: while each time capsule is a single configuration, worlds are finite volumes in configuration space, sets containing infinitely many configurations. In this respect, i.e., regarding their extension through configuration space, they more closely resemble Everettian world branches than Barbourian time capsules. However, they also differ from the former: Everettian branches do

sentient being can have the impression of perceiving the cup falling, records of the states in between. As we will see, the situation proves to be somewhat simpler for a model including universal time.

[35] Decoherence is a process that effectively suppresses interference effects via entanglement of a quantum system with its environment (Schlosshauer, 2014, p. 1): "Realistic quantum systems are never completely isolated from their environment. When a quantum system interacts with its environment, it will in general become entangled with a large number of environmental degrees of freedom. This entanglement influences what we can locally observe upon measuring the system. In particular, quantum interference effects with respect to certain physical quantities (most notably, 'classical' quantities such as position) become effectively suppressed, making them prohibitively difficult to observe in most cases of practical interest. This is the process of decoherence, sometimes also called dynamical decoherence or environment-induced decoherence. Stated in general and interpretation-neutral terms, decoherence describes how entangling interactions with the environment influence the statistics of results of future measurements on the system."

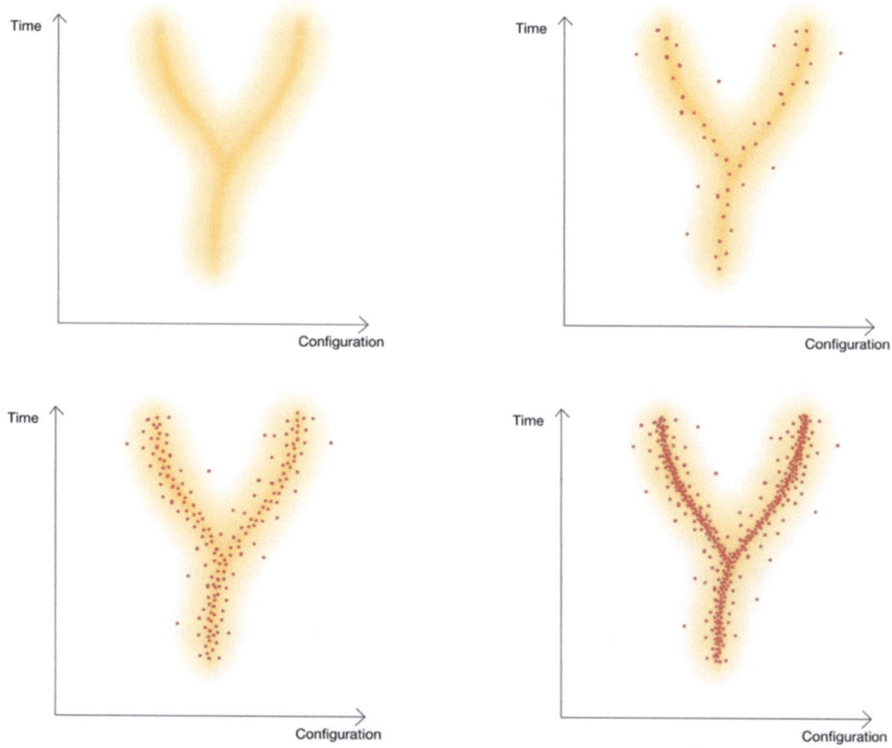

Fig. 2.1 (This figure first appeared in Dürr and Ehmann (2021)). The universal WF provides the probability density (yellow) for actualizations of configurations (red). Actualizations happen in accordance with the probability density, the super-particle thereby carving out the WF in configuration space. The latter decoheres into world branches. For illustrative purposes, only two branches are depicted. Depending on the actualization rate R, configurations form a more or less dense history. From top left to bottom right, increasing actualization rates are shown. In the limit of $R \to \infty$, every world branch is instantiated infinitely many times in any finite time interval. For suitable combinations of temporal coarse-graining and actualization rate, they quasi-persist

subsume histories of macroscopic change. Not so worlds. They, in themselves, are not histories of macroscopic change, since they only contain configurations that are macroscopically indistinguishable from a given macro state. In this respect, worlds are more like internally static time capsules. Nevertheless, we will see in a later chapter how world histories, i.e., successions of worlds, collectively instantiating macroscopic change, are formed from individual, internally unchanging worlds. Spelling out in greater detail how worlds are connected over extended time spans promises to be rewarding in several respects. For one, it will be beneficial to TBM and its acceptance as a viable alternative to BM and other non-relativistic quantum theories in that it provides a deeper, more thorough understanding of how world histories are formed from TBM's fundamentally discontinuous dynamics of randomly actualized configurations. For another, the kinship of worlds and time capsules, both being internally (quasi-)[36] static, promises greater compatibility with Barbour's theory.[37] At the same time, the investigation will help to formulate a constructive criticism of the latter, possibly contributing to its improvement. I think there is good reason to be critical of Barbour's claim that the perception of macroscopic change can emerge from one individual, static configuration. In fact, based on our best scientific understanding, one may doubt that any kind of experience of sentient beings is possible from within only one such configuration. The presumption is that, in order to generate experience, in particular the perception of change, macroscopically relevant physical change is necessary, at least for beings the psychological states of which are instantiated by physical processes. Of course, there is no conclusive argument providing proof that we are such beings. But why throw overboard our best biology and neuroscience if we can have the cake and eat it, too? The suggestion I am going to put forward might solve this issue: we can have both, internally quasi-static worlds and the relevant macroscopic change as well, through the emergence of world histories.

The question that needs to be answered in order to make this proposal work is the following: how can world histories comprising macroscopic change arise from discontinuous instantiations of individual, internally quasi-static worlds? To tackle this problem, I will now delve into a related issue within Boltzmannian statistical mechanics (BSM). There, the standard way of partitioning phase space into disjoint macro regions corresponding to different macro states has unexpected consequences: upon traversing boundaries between macro regions, the continuous evolution of the micro state leads to a discontinuous change of the macro properties instantiated by the system in question. In fact, within the standard framework of BSM, whenever macroscopic change is modelled, it is modelled as discontinuous. This is at odds with the expectation that the thermodynamic evolution of a system, if not always, at least sometimes is continuous, and that BSM's models should be able to reflect that. This, I call the discontinuity problem. In the following chapter, I will make explicit the conditions that apply to the usual way of partitioning phase space. After exposing the discontinuity problem in detail and presenting three arguments against sharp boundaries between macro regions in phase space, I discuss some possible solutions to the issue at hand. Amongst them, the conceptual shift from disjoint macro regions to overlapping worlds defined via a similarity relation shall turn out to be fruitful. The account thus developed—apart from some minor adaptations—almost straightforwardly carries over to the problem of the emergence of world histories in TBM, thereby hinting at a deeper connection between (the revised framework of) BSM and TBM, rooting from the shared concept of worlds and the coarse-graining relation between the realms of the micro- and macroscopic.

[36] The qualification 'quasi-'applies to the notion of worlds, since they do comprise different configurations, such that actualizing one configuration within a world and then another within the same world amounts to microscopic change, while, at the same time, no macroscopic change occurs. By contrast, time capsules are individual configurations, which renders them static simpliciter.

[37] At least, I see no a priori reason for Barbour to not coarse-grain configuration space into worlds.

References

Albert, D. Z. (1996). Elementary quantum metaphysics. In J. T. Cushing, A. Fine, & S. Goldstein (Eds.), *Bohmian mechanics and quantum theory: An appraisal*. Springer Science+Business Media.

Allori, V. (2015). Primitive ontology in a nutshell. *International Journal of Quantum Foundations, 1*(3), 107–122.

Allori, V., Goldstein, S., Tumulka, R., & Zanghì, N. (2008). On the common structure of Bohmian mechanics and the Ghirardi-Rimini-Weber theory. *The British Journal for the Philosophy of Science, 59*(3), 353–389.

Bacciagaluppi, G. (2016). rhe Role of Decoherence in quantum mechanics. In E. N. Zalta (Ed.). *The Stanford Encyclopedia of Philosophy* (Fall 2016 Edition). https://plato.stanford.edu/archives/fall2016/entries/qm-decoherence/

Barbour, J. (1999). *The end of time*. Oxford University Press.

Barnes, L. A. (2012). The fine-tuning of the universe for intelligent life. *Publications of the Astronomical Society of Australia, 29*(4), 529–564. https://doi.org/10.1071/AS12015

Barrett, J. A. (1996). Empirical adequacy and the availability of reliable records in quantum mechanics. *Philosophy of Science, 63*(1), 49–64. https://doi.org/10.1086/289893

Bell, J. S. (1987a). Quantum mechanics for cosmologists. In J. S. Bell (Ed.), 1987b *Speakable and unspeakable in quantum mechanics*. Cambridge University Press.

Bell, J. S. (1987b). *Speakable and unspeakable in quantum mechanics*. Cambridge University Press.

Bohm, D. (1952a). A suggested interpretation of the quantum theory in terms of “Hidden” variables. I. *Physical Review, 85*(2), 166–179. https://doi.org/10.1103/PhysRev.85.166

Bohm, D. (1952b). A Suggested interpretation of the quantum theory in terms of “Hidden” Variables. II. *Physical Review, 85*(2), 180–193. https://doi.org/10.1103/PhysRev.85.180

Bohm, D., & Hiley, B. J. (1993). *The undivided universe: An ontological interpretation of quantum theory*. Routledge.

Butterfield, J. (2002). Review: Julian Barbour, the end of time. *The British Journal for the Philosophy of Science, 53*(2), 289–330.

Callender, C. (2007). The emergence and interpretation of probability in Bohmian mechanics. *Studies in History and Philosophy of Modern Physics, 38*(2), 351–370. https://doi.org/10.1016/j.shpsb.2006.08.004

Carroll, S. M. (2017). Why Boltzmann Brains Are Bad. arXiv:1702.00850v1 [hep-th]

Chen, E. K. (2019). Realism about the wave function. *Philosophy Compass, 14*(7). https://doi.org/10.1111/phc3.12611

Cushing, J. T. (1994). *Quantum mechanics: historical contingency and the Copenhagen hegemony*. The University of Chicago Press.

Dürr, D., Goldstein, S., & Zanghì, N. (1992). Quantum equilibrium and the origin of absolute uncertainty. *Journal of Statistical Physics, 67*(5–6), 843–907. https://doi.org/10.1007/BF01049004

Dürr, D., Goldstein, S., & Zanghì, N. (2012). *Quantum physics without quantum philosophy*. Springer Science & Business Media.

Dürr, D., & Teufel, S. (2009). *Bohmian mechanics*. Springer Science & Business Media.

Dürr, P., & Ehmann, A. (2021). The physics and metaphysics of Tychistic Bohmian mechanics. *Studies in History and Philosophy of Science Part A, 90*, 168–183. https://doi.org/10.1016/j.shpsa.2021.09.014

Esfeld, M. (2014). Quantum humeanism, or: Physicalism without properties. *The Philosophical Quarterly, 64*(256), 453–470. https://doi.org/10.1093/pq/pqu030

Esfeld, M. (2018). Collapse or No Collapse? What Is the best ontology of quantum mechanics in the primitive ontology framework? In S. Gao (Ed.), *Collapse of the wave function*. Cambridge University Press.

Frege, G. (1892). *Über Sinn und Bedeutung*. Zeitschrift für Philosophie und philosophische Kritik, Neue Folge, Band 100/1, 25-50. Pfeffer

Friebe, C., Kuhlmann, M., Lyre, H., Näger, P., Passon, O., & Stöckler, M. (2015). *Philosophie der Quantenphysik*. Springer-Verlag. https://doi.org/10.1007/978-3-642-37790-7

Friederich, S. (2018). Fine-Tuning. In E. N. Zalta (Ed.). *The Stanford Encyclopedia of Philosophy* (Winter 2018 Edition). https://plato.stanford.edu/archives/win2018/entries/fine-tuning/

Goldstein, S. (2012). Typicality and notions of probability in physics. In Y. Ben-Menahem & M. Hemmo (Eds.), *Probability in physics*. Springer Science & Business Media.

Goldstein, S. (2017). Bohmian mechanics. In E. N. Zalta (Ed.). *The Stanford Encyclopedia of Philosophy* (Summer 2017 Edition). https://plato.stanford.edu/archives/sum2017/entries/qm-bohm/

Goldstein, S., Huse, D. A., Lebowitz, J. L., & Tumulka, R. (2017). Macroscopic and microscopic thermal equilibrium. *Annalen der Physik, 529*, 1600301. https://doi.org/10.1002/andp.201600301

Goldstein, S., Tumulka, R., & Zanghì, N. (2011). Bohmian trajectories as the foundation of quantum mechanics. In P. K. Chattaraj (Ed.), *Quantum trajectories*. CRC Press.

Hacking, I. (1987). The Inverse Gambler's fallacy: The argument from design. The anthropic principle applied to wheeler universes. *Mind, XCVI*(383), 331–340.

Hiley, B. J. (1999). Active information and teleportation. In D. Greenberger, W. L. Reiter, & A. Zeilinger (Eds.), *Epistemological and experimental perspectives on quantum physics*. Springer Science & Business Media.

Hiley, B. J. (2009). Bohm interpretation of quantum mechanics. In D. Greenberger, K. Hentschel, & F. Weinert (Eds.), *Compendium of quantum physics*. Springer Science & Business Media.

Maudlin, T. (1995). Three measurement problems. *Topoi, 14*(1), 7–15. https://doi.org/10.1007/BF00763473

McLaughlin, B., & Bennett, K. (2018). Supervenience. In: Zalta, E. N. (ed.). *The Stanford Encyclopedia of Philosophy* (Winter 2018 Edition). https://plato.stanford.edu/archives/win2018/entries/supervenience/

Ney, A. (2012). The status of our ordinary three dimensions in a quantum universe. *Noûs, 46*(3), 525–560. https://doi.org/10.1111/j.1468-0068.2010.00797.x

Norsen, T. (2018). On the explanation of born-rule statistics in the de Broglie-Bohm Pilot-Wave theory. *Entropy, 20*(6), 422–426. https://doi.org/10.3390/e20060422

Norton, J. D. (2015). *You are not a Boltzmann Brain*. Retrieved January 25, 2018, from https://www.pitt.edu/~jdnorton/Goodies/Boltzmann_Brain/Boltzmann_Brain.html

Oldofredi, A., & Esfeld, M. A. (2018). *On the possibility of a realist ontological commitment in quantum mechanics*. arXiv:1801.05307

Oppenheimer, D. M., & Monin, B. (2009). The retrospective gambler's fallacy: Unlikely events, constructing the past, and multiple universes. *Judgement and Decision Making, 4*(5), 326–334.

Passon, O. (2010). *Bohmsche Mechanik*. Europa-Lehrmittel.

Rowbottom, D. (2015). *Probability*. Polity Press.

Schlosshauer, M. (2007). *Decoherence and the quantum-to-classical transition*. Springer.

Schlosshauer, M. (2014). *The quantum-to-classical transition and Decoherence*. https://arxiv.org/abs/1404.2635v2

Suárez, M. (2015). Bohmian dispositions. *Synthese, 192*(10), 3203–3228. https://doi.org/10.1007/s11229-015-0741-1

Tahko, T. E. (2018). Fundamentality. In: E. N. Zalta (Ed.). *The Stanford Encyclopedia of Philosophy* (Fall 2018 Edition). https://plato.stanford.edu/archives/fall2018/entries/fundamentality/

Valentini, A. (1991). Signal-locality, uncertainty, and the subquantum H-theorem. I. *Physics Letters A, 156*(1,2), 5–11. https://doi.org/10.1016/0375-9601(91)90116-P

Wallace, D. (2012). *The emergent multiverse*. Oxford University Press.

Chapter 3
Smoothing out Boltzmannian Statistical Mechanics: On the Viability of an Alternative Partitioning of Phase Space Into Overlapping Worlds

3.1 Reducing Thermodynamics to Boltzmannian Statistical Mechanics: The Case of Macro Values[1]

A core method of the framework of Boltzmannian Statistical Mechanics (BSM) is the partitioning of phase space into non-overlapping macro regions, i.e., disjoint sets of micro states. Macro regions are of finite, non-zero volume, and stand in a one-to-one correspondence with different macro states.[2] Given a finite volume of phase space to be partitioned, there is a finite number of macro regions in a partition. Macro states are specified via particular macro values, e.g., for temperature, volume and pressure of a gas. In general, macro states are defined and specified via particular macro values taken on by macro variables: following Werndl and Frigg (2015b, p. 1225), a system "can be characterized by a set $\{v_1,\ldots,v_k\}$ of *macrovariables* $(k \in \mathbb{N})$. The v_i assume values in [macro value space] $\mathbb{V}_i$, and capital letters V_i denote the values of v_i. A particular set of values $\{V_1,\ldots,V_k\}$ defines a *macro-state* $M_{V_1,\ldots,V_k}$."

Boundaries between macro regions are construed as being sharp: under a given partition, each micro state is a member of one and only one macro region. Since macro regions stand in a one-to-one correspondence with macro states—i.e., there

---

[1] This paragraph, modulo some small adaptations, has recently been published with Synthese, see Ehmann (2022). Reproduced with permission.

[2] See, e.g., Callender (1999), Frigg and Werndl (2012), Werndl and Frigg (2015b), Goldstein et al. (2017). The following discussion is based mainly on the framework of BSM as presented in the writings of Charlotte Werndl and Roman Frigg, but also others. I will often refer to it as the "usual framework", for it is fairly common, without intending to imply that there aren't any alternatives. Thus, the issues identified pertain at least to this "usual" account, but not necessarily to all possible accounts. In particular, I deem it possible to avoid the discontinuity problem described below by adopting a different way of carving up phase space. Such an alternative I will present later.

A. Ehmann, *Worlds in a Stochastic Universe: On the Emergence of World Histories in Tychistic Bohmian Mechanics*, Synthese Library 537,
https://doi.org/10.1007/978-3-032-14632-8_3

is a bijection between the set of macro regions and the set of macro states—each micro state instantiates one and only one macro state, and micro states from different macro regions instantiate different macro states. Also, once fixed, a given partition doesn't change (Hemmo & Shenker, 2012, p. 53): "The partition of the state space into macrostate regions is fixed in time"; and (ibid., 56): "the partition of the state space into macrostates [...] is time-independent". This partition of phase space into a finite number of macro regions with sharp, fixed boundaries results in models that are unable to describe continuous macroscopic change: whenever macroscopic change is modelled within the framework of BSM, it is modelled as discontinuous. That is, according to BSM's models, every time the micro state of a system crosses the boundary between two macro regions, a discontinuity in (at least one of) the system's macro values results.

Conceptually, this is at odds with the thermodynamic description of the continuous time-evolution of macro values, and likewise with the continuous evolution of the micro state in phase space. From this discontinuity problem (DP), as I call it, ensues a problem for the foundational project[3] of reducing thermodynamics to BSM.[4] The foundational project is, as Callender (1999, p. 348) puts it, "the project of demonstrating how real mechanical systems can behave thermodynamically."

Take, for example, a gas consisting of a number N of particles and assume that classical mechanics provides a correct description of the fundamental goings-on, i.e., of the dynamics of the N particles. The micro state of such an N-particle system is represented by a point x in $6N$-dimensional phase space Γ, specifying all positions and momenta of the N particles. "As the system evolves through time, this representative point will trace out a trajectory through Γ. We thus have two descriptions of our gas: one mechanical and the other thermodynamical. We would like to know how they relate to one another. This kind of problem is quite familiar to philosophers. Philosophers of biology, psychology, and all of the special sciences are busy trying to demonstrate how the properties and concepts used by their science are (not) reducible in some sense to the properties and concepts of 'lower level' sciences. The problem is one of intertheoretic reduction." (Callender, 1999, p. 350f.)

The thrust of this paragraph is to show that this inter-theoretic reduction of thermodynamics to BSM isn't as straightforward as it might seem at first—even in the seemingly simple case of reducing the thermodynamic concept of temperature to its analogue in BSM—and to show why this is so. This problem will be framed in terms of models and their (in)adequate tracking of their target systems. The question to be answered, then, is the following: Is the evolution of macro values of thermodynamic target systems adequately tracked by BSM's models? It will be answered in the negative: the usual BSM framework, in principle, doesn't allow for models that adequately track the evolution of thermodynamic macro values. This,

[3] I adopt the terminology from Callender (1999, p. 348).

[4] Throughout this book, I adopt the "Philosopher's sense of reduction", see Batterman (2020).

while being unproblematic for the practitioner, constitutes a conceptual hurdle for the foundational project.[5]

Many of the literature on the reduction of thermodynamics to statistical mechanics engages with the tension between time-reversal invariant micro dynamics and the second law of thermodynamics. This is not the subject of the present investigation. The problem at hand, in a sense, is more basic: while thermodynamic descriptions allow for continuous change of temperature macro values, descriptions provided by BSM's models don't. It is in this sense that the latter do not adequately track the former. The two descriptions come apart: macroscopic temperature, as described by BSM, behaves differently in its evolution over time than macroscopic temperature as described by thermodynamics. This indeed is puzzling, and poses a threat to the foundational project: with macroscopic temperature in thermodynamics (T_{TD}) differing in this way from macroscopic temperature in BSM (T_{BSM}), the possibility of reducing one to the other is put into question. At the very least, T_{TD} and T_{BSM} aren't readily identified with one another, as might seem possible at first sight. What's more, as will become evident later, T_{TD} doesn't supervene on T_{BSM}, a condition generally regarded as a necessary requirement for reduction. As van Riel & Van Gulick (2019, 4.5.3) write: "it [supervenience] surely is a necessary condition for reduction". Likewise, McLaughlin & Bennett (2018, 3.3): "Everyone agrees that reduction requires supervenience". But with supervenience being a *conditio sine qua non* for reduction, and T_{TD} not supervening on T_{BSM}, the former cannot be reduced to the latter.[6]

Models describing change of macro values as always discontinuous are at odds with thermodynamic descriptions of these macro values: a system's macroscopic evolution as described by thermodynamics involves continuous change of its macro properties—of the temperature value, for example. Likewise, such models are at odds with the continuous evolution of the micro state x through Γ. A continuous evolution of the micro state should not lead to discontinuous change of the macro state, i.e., should not lead to a discontinuity in the evolution of macro values. Yet, as the framework of BSM is set up, continuous change of the micro state sometimes (i.e., upon traversing boundaries between macro regions) *does* lead to discontinuous change of the macro state. In fact, the models resulting from the framework of BSM predict that, *whenever* there is macroscopic change, it is sudden and of a jumpy nature: during its wanderings *within* a macro region, the micro state always and invariably instantiates the same, unchanging macro state, according to the

[5] Like Callender's (see ibid., 349), my quarrels are philosophical in nature, concerning the foundational project. The practicing scientist needn't be concerned.

[6] My aim here is *not* to argue against the foundational project of reducing thermodynamics to BSM, or to attack the latter. Rather, the goal is to contribute to the foundational project by providing counterexamples that show where BSM and its models, as it stands, have their shortcomings, and how these threaten the foundational project. Following (Wimsatt, 2006, p. 473), "counterexamples become revealing sources of information about limitations of a model, or suggestions for probing its depths; in either case, a tool to refine the model, not an argument for trashing the system, or something to be swept under the rug."

framework's models. As soon as the micro state moves into a *different* macro region, however, it suddenly instantiates a different macro state: since boundaries between macro regions are construed as sharp, macro values jump upon traversal of the micro state from one macro region to another. This results from the definition of macro states via "particular set[s] of [macro] values", and the bijective association of the so defined macro states with macro regions of finite volume (Werndl & Frigg, 2015b, p. 1225): "a system's microstate uniquely determines its macrostate. Every macrostate M is associated with a macroregion Γ_M consisting of all $x \in \Gamma_E$ [with Γ_E denoting the energy hypersurface, i.e., the complete available phase space for a constant energy] for which the system is in M." The consequence of this construal is that at least one macro value always jumps upon the micro state crossing boundaries between macro regions. Assuming that the continuous thermodynamic evolution of macro states is to be recovered when reducing thermodynamics to BSM, jumping macro values, especially upon tiny changes of the micro state, are ill-suited. Yet, the partition of phase space into disjoint macro regions of finite, non-zero volume with fixed, sharp boundaries, as prescribed by the usual framework, results in models describing jumping macro states.

This discontinuity problem, merely sketched so far, will be discussed in detail in the following. At this point, some emphasis should be put on the fact that DP only pertains to models resulting from the usual framework of BSM, and explicitly not to the target systems described by those models. The claim is neither that the physical systems themselves suffer from DP, i.e., that they always undergo discontinuous macroscopic change, if they undergo macroscopic change at all. Nor, that the thermodynamic description of these physical systems suffers from DP. Rather, only the BSM models that describe them entail these unavoidable discontinuities. This, in itself, would even be unproblematic. Per se, there is nothing wrong about models allowing only for discontinuous change. DP arises only if one wants to have these models adequately describe systems that undergo continuous macroscopic change as described by thermodynamics. Hence, what renders DP problematic is the discrepancy between two modes of description—BSM's models on the one hand and the thermodynamic description on the other. And this discrepancy becomes problematic in particular within the context of the foundational project. However, if no inter-theoretic reduction of thermodynamics on BSM is sought, and in particular, if no reduction of T_{TD} on T_{BSM} is sought, one might as well live happily with this discrepancy.

3.1.1 *The Framework of BSM*

Let's take a closer look at the framework of BSM as it is usually presented, e.g., in discussions of approach to equilibrium[7]: We are interested in a (classical) system comprised of N constituents that move through physical space. The micro state of

[7] Roughly following the presentation in Frigg and Werndl (2012, p. 101f).

the system—a complete specification of the positions and momenta of all the constituents involved in the system—is represented by a point x in its $6N$-dimensional phase space Γ. (If the energy of the system is constant, the phase space volume available to the system is reduced to a $(6N-1)$-dimensional energy hypersurface Γ_E.) The available phase space is partitioned into sub-regions that correspond to different macro states M_i—specified by particular macro values taken on by macro variables, see above—of the system: the sub-region in Γ corresponding to a macro state M_i is the set of all $x \in \Gamma$ that instantiate M_i and called the macro region Γ_{M_i}. Macro regions Γ_{M_i} form a partition of Γ: they cover all of the available Γ and do not overlap. A macro state of special interest is M_{eq}, the equilibrium state, instantiated by any micro state $x \in \Gamma_{M_{eq}}$ (and counterfactually instantiated[8] by all micro states $x \in \Gamma_{M_{eq}}$). Given the standard Lebesgue measure, the volume of sub-regions of Γ can be assessed and hence is rendered comparable. It is usually said that (Frigg & Werndl, 2012, p. 102) "for gases $\Gamma_{M_{eq}}$ is vastly larger […] than any other macro-region, a fact also known as the 'dominance of the equilibrium macrostate' […]; in fact Γ_E is almost entirely taken up by equilibrium microstates. For this reason the equilibrium state has maximum entropy."

3.1.2 *Partitioning Phase Space into Macro Regions*

Of special importance for the purposes of this chapter is the notion of macro regions and their partitioning phase space (ibid.): "From a macroscopic perspective, the system is characterised by a set of *macrostates* $M_i, i = 1, \ldots, m$. To each macrostate corresponds a macro-region Γ_{M_i} consisting of all $x \in \Gamma_E$ for which the system is in M_i. The Γ_{M_i} form a partition of Γ_E, meaning that they do not overlap and jointly cover Γ_E." Frigg et al. (2016, 4.1) summarise the key feature of the correspondence between micro states and macro states in a short statement: "to every given microstate x there corresponds *exactly one* macrostate".

For one thing, this means that x either instantiates a certain macro state M_i or another macro state M_j, $(i \neq j)$, but never both. Since macro states M_i and M_j correspond to macro regions Γ_{M_i} and Γ_{M_j}, respectively, x must not belong to two macro regions. This can be called the *no-overlap* condition:

$$\forall x \forall \Gamma_{M_{i,j}} \neg\left(x \in \Gamma_{M_i} \wedge x \in \Gamma_{M_j}\right); i \neq j \qquad \text{(no-overlap)}$$

For another, it means that macro regions exhaustively cover the available phase-space volume. All micro states must instantiate *exactly one* macro state, read as:

---

[8] "Counterfactually instantiated" in the sense that at any time, only one micro state is actualized, and thus only one micro state actually instantiates the macro state, but the actualization of the other micro states in the same macro region *would* instantiate the same macro state, if they *were* actualized instead.

every micro state must instantiate *some* macro state, and hence belong to *some* macro region. It cannot belong to *no* macro region at all. Hence, all of the available phase space must be covered by macro regions ("they … jointly cover Γ_E." (Frigg & Werndl, 2012, 102.)) This can be called the *exhaustive-cover* condition:

$$\forall x \exists \Gamma_{M_i} \left(x \in \Gamma_{M_i} \right) \qquad \text{(exhaustive-cover)}$$

Taken together, *no-overlap* and *exhaustive-cover* ensure that there is always some macro state instantiated, that this macro state is internally consistent, i.e., that a micro state never instantiates two (or more) distinct macro states at the same time, and that macro regions are not fuzzy: while it might be the case that there is some uncertainty or even complete ignorance about which macro state a given system is actually in—if only because the respective macro values haven't been measured/observed—one can always be sure that its actual micro state x is located in *some* macro region or other, i.e., in one (*exhaustive-cover*) and only one (*no-overlap*) macro region Γ_{M_i}, and that thus, one and only one macro state is instantiated at any given time. After all, fuzzy boundaries would mean that the actual micro state does not always belong to a certain (i.e., to one and only one) macro region. Being located at the fuzzy boundary, x could belong to Γ_{M_i}, to Γ_{M_j}, to neither, or to both. The latter option "both" is excluded by *no-overlap*. The option "neither" is excluded by *exhaustive-cover*. Hence, it must belong to either one of the former. In other words: boundaries of macro regions are sharp.

These considerations about the overlap and fuzziness of macro regions—or rather: these stipulations of their impossibility—are reflected in the schematic pictorial representations of phase-space partitioning we find, e.g., in Goldstein et al. (2019), (Fig. 3.1)[9]. As can be seen, macro regions cover all of phase space, do not overlap and have sharp boundaries.

3.1.3 The Origin of the Discontinuity Problem

As stated above, macro regions in phase space are of finite, non-zero volume, and there is a finite number of macro regions in a partition. Why is that so? First, let's answer the question why macro regions must have a finite, non-zero volume. The reason for this lies in the way BSM works. The Boltzmann entropy function—$S_B(M_i) := k_B \log\left[\mu\left(\Gamma_{M_i}\right)\right]$[10]—relies on the volume of macro regions in order to assign different entropy values to different macro states of a system. If,

[9] I thank Sheldon Goldstein and the publisher of Annalen der Physik for the permission to reproduce this figure. It first appeared in Goldstein et al. (2017). Copyright Wiley-VCH Verlag GmbH & Co. KGaA. Reproduced with permission.

[10] See Frigg and Werndl (2012, p. 102), Werndl and Frigg (2015b, p. 1225).

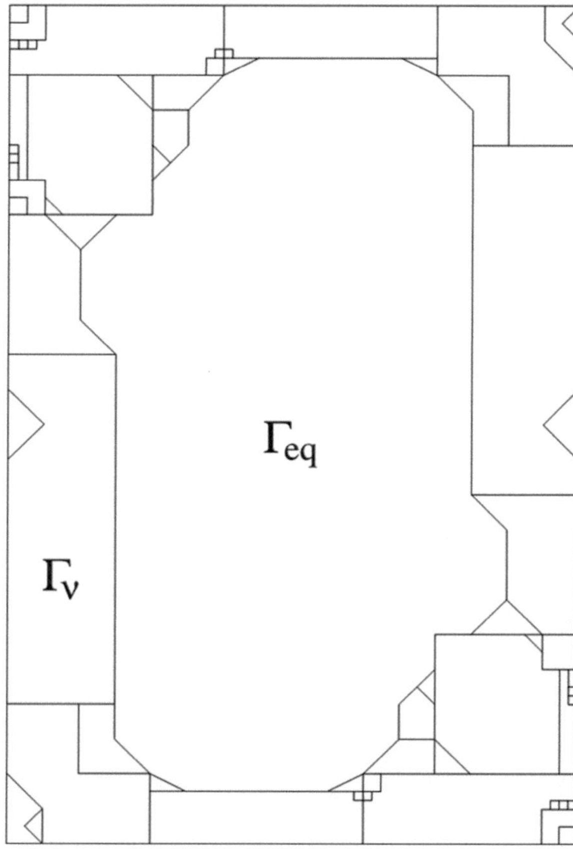

Fig. 3.1 A partition of phase space into macro regions

contrary to this, macro regions in fact were of volume zero, the entropy function would assign to macro states no well-defined entropy value. This would seriously undermine the spirit and functionality of BSM. For example, it would be impossible to mark the equilibrium macro state as dominant in virtue of its having the highest entropy. Likewise, arguments about the relation of entropy change and the arrow of time would be blocked. This consequence is to be avoided, if only for the fact that it would be too high of a price to pay in comparison to what potentially could be gained by having zero volume macro regions. In order to retain its spirit and functionality, BSM must construe macro regions as having finite volume greater zero.

Let's move on to the second part of the statement above: there is a finite number of macro regions in a partition. The reason for that is rather simple. Given macro regions, as construed within the framework of BSM, are of finite, non-zero volume (and do not approach zero volume), it is trivial that the finite volume of phase space available to a system can only be partitioned into a finite number of macro regions. But why even assume that the available phase space volume is finite? If it were infinite, it could be partitioned into an infinite number of compartments, despite the latter having a finite, non-zero volume. However, it is unreasonable to assume an infinite phase space volume for at least many of the systems BSM is supposed to model. For example, in the paradigmatic gas in a box, the spatial degrees of freedom each constituent has are bounded: particles can only move within the box (itself of finite volume). Such cases should be covered by BSM, so it is uncalled for to simply assume that BSM only models systems that are not limited to a finite spatial volume. Likewise, for finite temperatures, the momenta of the particles can only take on values within certain limits. Accordingly, the phase space volume of the entire system is restricted to be finite, if very large. And once the available phase space volume is finite, it is impossible to partition it into an infinite number of compartments with finite, non-zero volume. So the number of macro regions in a partition must be finite.

Having established that there is a finite number of macro regions in a partition, there can only be a finite number of different macro states. Macro states and macro regions, as described before, stand in a one-to-one correspondence. That is, the mapping between macro regions and macro states is bijective. But it is impossible to map a set containing an infinity of elements—let alone a set containing an uncountable infinity of elements—bijectively onto a set containing only a finite number of elements. Accordingly, it is impossible to bijectively map infinitely many macro states onto finitely many macro regions. So, if there is a finite number of macro regions, they can only correspond to a finite number of different macro states. In fact, the number of macro states and the number of macro regions in the model must be equal: there are precisely as many macro states as there are macro regions, and this number is finite.

Given there is a finite number of different macro states, there can only be a finite number of different sets of macro values corresponding to them. In the BSM framework, a macro state, as defined above, is specified by a set of macro variables, taking on macro values. The simplest case, of course, is if this set only contains one relevant element, such that we can ignore the rest. Take, for example, some macro states that differ only in their temperature macro values. In this case, we can ignore other macro variables: they are assumed to take on identical values for the different macro states. Since there is a finite number of different macro states in the model, and these macro states differ in their temperature macro value, there is a finite number of different temperature macro values. The reason for this is similar to the one just discussed in the case of macro regions and macro states: the set of different macro states, since the latter are specified by temperature macro values, stands in a one-to-one correspondence to the set of different temperature macro values. So with a finite number of different macro states, there is a finite number of different temperature macro values specifying the former.

It is important to recognize that *different* macro regions must correspond to *different* macro states, and thus to *different* macro values. For if different macro regions would correspond to the same macro state, i.e., to identical macro values, they wouldn't be different macro regions in the first place. Such macro regions, corresponding to the same macro state, would be combined into one macro region. This is a consequence of the one-to-one correspondence between macro regions and macro states in the framework of BSM. In effect, different macro regions must correspond to different macro states, and, a fortiori, to different macro values. In the simplified example, being concerned only about the temperature macro variable, this means that different macro regions must correspond to different temperature macro values. But, as argued above, there is only a finite set of macro regions, and hence a finite set of different temperature macro values assigned to them. What, then, does it mean for a finite set of temperature macro values to consist of *different* temperature macro values? It means that, when subtracting any value from this set from any other, the result must always be different from zero, for if it was zero, the two values would be identical. It is unreasonable to assign identical macro states, i.e., macro states not differing in their macro values, to different macro regions, as argued before: those macro regions would be combined into one macro region, such that, in the end, macro regions and macro states remain in a one-to-one correspondence, and likewise macro regions and (sets) of macro values.

So, there is a finite set of different macro values, each specifying a different macro state, corresponding to different macro regions. But if all pairs of elements of the set of macro values are different in that the subtraction of any one of them from any other results in a non-zero difference, then the transition between two macro values, i.e., between two macro states, i.e., between two macro regions, is always modelled as discontinuous. To illustrate this, imagine the temperature macro value of a system as modelled by BSM graphed as a function over time (Fig. 3.2). As long as the micro state wanders through a macro region Γ_{M_i}, the temperature macro value $T(t) = T_{M_i}$ remains absolutely constant, according to the model. So the graph shows a continuous, straight line parallel to the time axis, representing the constant temperature macro value. But at the precise instant the micro state crosses the boundary and moves into another macro region Γ_{M_j}—call this instant $t_{crossing}$—the temperature macro value changes to $T(t) = T_{M_j}$. Then, as long as the micro state stays in this macro region, the temperature macro value again remains absolutely

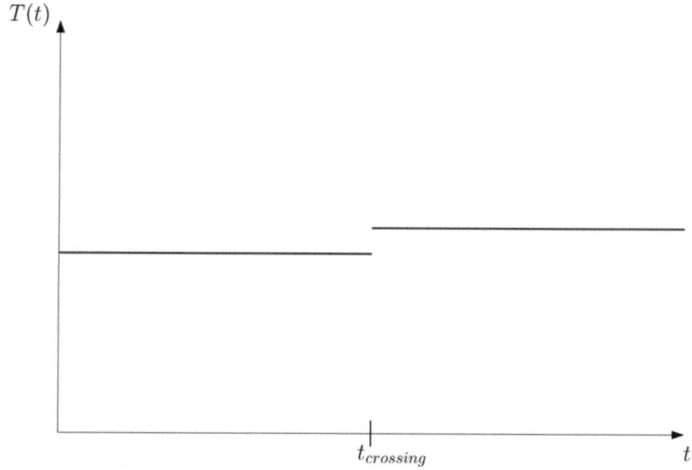

Fig. 3.2 A jump discontinuity in the temperature value

constant, so the graph again shows a continuous, straight line parallel to the time axis, representing this different temperature macro value. Thus, the graphed function, at $t_{crossing}$, contains a discontinuity which is not removable: consider the limit $L^- = \lim_{t \to t^-_{crossing}} T(t)$ of the function for t approaching $t_{crossing}$ from the left side, i.e., for the time during which the micro state moves through the first macro region Γ_{M_i}. The value of this limit is the temperature value before crossing the boundary: $L^- = T_{M_i}$. Now consider the limit $L^+ = \lim_{t \to t^+_{crossing}} T(t)$ of the function for t approaching $t_{crossing}$ from the right side, i.e., for the time during which the micro state moves through the second macro region Γ_{M_j}, traced backwards. The value of this limit is the temperature value after crossing the boundary, $L^+ = T_{M_j}$. Since $T_{M_i} \neq T_{M_j}$, the limit from the left side and the limit from the right side are unequal as well: $L^- \neq L^+$. So there exists no single limit; the discontinuity at $t_{crossing}$ indeed is a jump discontinuity.

This is how the discontinuity problem arises from the usual framework of BSM. As stated before, DP consists in the discrepancy between the thermodynamic description of the time evolution of macro values, which can change continuously, and the fact that, in the usual framework of BSM, continuous change of macro values can't be modelled. So it neither consists solely in the fact that BSM is unable to model continuous change alone, nor does it contain the claim that in nature, or in thermodynamics, or in reality, macroscopic change is always discontinuous. The discontinuity encountered in BSM's models is an artefact resulting from the way the framework is set up: partitioning phase space into a finite number of macro regions, standing in a one-to-one correspondence with different macro states, the latter themselves standing in a one-to-one correspondence with different sets of macro values specifying them, such that, in the end, the elements of a finite set of macro regions stand in a one-to-one correspondence with the elements of a finite set of macro values— this partitioning, judged from the thermodynamic viewpoint, where continuous change of macro values is allowable, is an idealisation. In which sense this idealisation is problematic will be discussed in the following.

3.1.4 Models, Fundamentality, and the Discontinuity Problem

The discontinuity problem of BSM is put into sharp relief in the context of the foundational project of reducing thermodynamics to BSM, regarding BSM as the fundamental theory, as for example Frigg and Werndl (2019) do. They classify BSM as a fundamental, true and complete theory (within the scope of statistical mechanics, and as opposed to Gibbsian statistical mechanics (GSM), which they consider to be an effective theory): "BSM is a fundamental theory" (425, 430, 436); "BSM provides a true description of the systems within the scope of SM" (425); "BSM provides the complete fundamental theory of SM systems" (431). By contrast to effective theories, fundamental theories are supposed to provide correct descriptions of the world (ibid. 431): "BSM is quite unlike GSM [...]. Dynamical considerations occupy centre stage in BSM. It introduces macrostates with corresponding macroregions, and then defines equilibrium in explicitly dynamical terms (namely as the macrostate whose macroregion is such that, in the long run, the system's state spends most of its time in that macroregion). [...] [U]nder the assumption that the world is governed by Newton's equation of motion [...] the dynamics considered in BSM is the true dynamics at the fundamental level: the unabridged and unidealised dynamics with all interactions between all microconstituents of the system. Equilibrium results from macrostates that are defined in terms of macrovariables that supervene on the true microdynamics of the system, and where a system fluctuates away from equilibrium it does so as a result of the true underlying dynamics. In a classical world the theory gives a full account of all this – nothing is left out and nothing is averaged over. BSM provides the complete fundamental theory of SM systems."

Clearly, Frigg and Werndl engage in the foundational project, explaining macroscopic (thermodynamic) behaviour qua reduction to the more fundamental level as modelled by BSM. In order to make clear why this isn't as trivial as it seems, one first needs to distinguish between two types of functions models can have, and the way in which they can be deficient.

The first type of function of models is to "*represent* a selected part or aspect of the world, which is the model's target system." (Frigg & Hartmann, 2020, 1., my emphasis.) Scale models have this function. They are (ibid.) "down-sized or enlarged copies of their target systems. A typical example is a small wooden car that is put into a wind tunnel to explore the actual car's aerodynamic properties." Whether a model fulfils its representational function successfully depends on the application. The wooden car model might be successful in representing the aerodynamic aspects of the real car, but it certainly fails in representing the real car's material composition (see ibid.). It is deficient in the sense that it doesn't represent all aspects of the real car. With respect to the intended application, this deficiency isn't problematic. There, its function is to represent a certain aspect of the real car, not all of them.

A different type of function of models is to *describe* a target system that falls within the domain of description of a given theory or theoretical framework. For example, a system of two massive objects orbiting each other in empty space is a target system that falls within the domain of description of Newtonian classical mechanics. Models describing a target system often entail approximations or idealisations. Borrowing the distinction from Norton (2012, p. 210), "an approximation is an inexact description of a target system." For example, using Newtonian mechanics to describe the target system of two objects orbiting each other in empty space results in a model ignoring relativistic effects. This amounts to an approximation. "An idealization", on the other hand, "is a real or fictitious system, distinct from the target system, some of whose properties provide an inexact description of some aspects of the target system" (ibid.). (Fictitiously) modelling the above two-body target system as a system of two point-like masses would be an idealization. This model isn't merely an approximation. It also has a representational function in that it represents the aspects "mass" and "position" of the target system in a fictitious system of point-like masses. Different functions of models aren't necessarily mutually exclusive. Approximations are deficient in virtue of being inexact, and the same is true for idealisations, for they entail approximations. Idealisations however, because they are systems themselves (if only fictitious ones), may also be deficient in virtue of not representing all aspects of the target system. Again, these deficiencies aren't problematic, as long as the model fulfils its intended function.

The function of BSM's models is of the second type. They describe some aspects of a target system, e.g., of a gas in a box, that falls within the domain of description of BSM. To be more precise, they describe the macro evolution of a thermodynamic target system, e.g., during approach to equilibrium. Due to its method of partitioning phase space, BSM's models are an idealisation in that they replace the thermodynamic target system with a fictitious system that can only instantiate a subset of the macro states the target system can assume.[11] Their representation of the time evolution of the target system, in other words, is deficient in virtue of not representing all aspects of it. This idealisation entails also an approximation: BSM's models ignore the actual macro values of the thermodynamic target system's macro variables at any given time and approximate them by assigning fixed macro values for finite intervals of time.

Now this *approximation* is unproblematic in practice. If the model is too crude, if the approximation is too inexact, one can always impose a more fine-grained partition. However, that BSM's models are *idealisations* with respect to the possible macro states a thermodynamic target system can take on *is* problematic for the foundational project. Working "under the assumption that the world is governed by Newton's equation of motion", Frigg & Werndl regard BSM as a "true complete fundamental theory". As Norton (2012) points out, one would expect the less fundamental theory to be an idealisation of the fundamental theory, not the other way around, as is the case here. Clearly, BSM's models being idealisations with respect to the possible macro states a system can take on is an oddity that must not be overlooked when engaging with the foundational project of reducing thermodynamics to BSM, especially if the latter is supposed to be not only a fundamental, but also a true and complete theory.

[11] By contrast, Frigg and Werndl (2019, p. 431) are correct when they write that the dynamics of the micro state "is the true dynamics at the fundamental level: the unabridged and unidealised dynamics with all interactions between all microconstituents of the system", given "the assumption that the world is governed by Newton's equation of motion". Reduction has already taken place "in the background" (Sklar, 1993, p. 348), in the sense that, as Sklar puts it (ibid.), "we have already identified gases as collections of interacting molecules and reduced our gas theory to a theory of molecules and their interaction." The threat to the foundational project doesn't consist in putting into question this background reduction. (More on that below.) That is, it doesn't stem from the ontology and micro dynamics assumed by BSM, which, in this context, i.e., given the above assumption, isn't an idealisation in Norton's sense. Rather, the threat exists due to discrepancies regarding macro properties and their evolution, and stems from BSM's method of partitioning phase space.

3.1.5 Partitioning Phase Space: A Known Issue

As described above, the construal of phase space partitioning within the framework of BSM, with sharp boundaries between a finite number of non-overlapping macro regions of finite, non-zero volume, underlies the discontinuity problem. At the same time, this construal seems to be standard. There are hints to be found in the literature, showing that not everyone considers it to be entirely unproblematic. Goldstein et al. (2019, 4) remark that phase space partitioning is not as straightforward a procedure as one might think: "$\Gamma(X)$ is the set of all phase points that 'look macroscopically the same' as X. Obviously, there is no unique precise definition for 'looking macroscopically the same,' so we have a certain freedom to make a reasonable choice". And later (ibid., 28), introducing the idea of "'fuzzy' macro sets", i.e., fuzzy macro regions: "The point here is to get rid of the sharp boundaries between [macro regions] as the boundaries are artificial and somewhat arbitrary anyway." I agree with Goldstein et al.'s verdict that boundaries are "artificial". They are an artefact, resulting from the way the framework of BSM is set up.[12]

Here is Wallace (2018, p. 19) taking a similar line regarding the arbitrariness of phase space partitioning in BSM: "the macrostate partition at the heart of Boltzmannian statistical mechanics is [...] vulnerable to these criticisms [(concerning the adequacy of coarse-graining)] [...] Consider some standard descriptions of the coarse-graining: '[W]e must partition [phase space] into compartments such that all of the microstates X in a compartment are macroscopically indistinguishable[.]' (Callender, 1999, p. 355). 'Everyday macroscopic human language (that is) carves the phase space of the universe up into chunks.' (Albert, 2000, p. 47) If pushed, I suspect Boltzmannians would reply that it is not the epistemic indistinguishability of macrostates that is doing the work, but rather the possibility of writing down robust higher-level dynamics in terms of macrostates, and largely abstracting over microscopic details". Wallace's argument here is a tu quoque: if the Gibbsian approach can rightly be criticized for coarse-graining, then the Boltzmannian

[12] That they are artificial is not to say that they are necessarily "subjective" or "anthropocentric". There is a strong analogy between coarse-graining in GSM and partitioning in BSM. Robertson (2020) defends both against attacks referring to the alleged subjectivity and anthropocentricity, justified by what Robertson (op.cit. 564) calls (ibid.) "the MI justification"—the appeal "to our [limited] observational capacities". As she points out, "the MI justification is unsatisfactory". Most importantly, without going into too many details, coarse-graining, and by analogy, partitioning, are (op.cit, 565) "abstraction[s] to a higher level of description", a level of description that in its own right is endowed with explanatory power and describes regularities in the behaviour of macroscopic systems such as gases (see op.cit. 566). All this would be lost if it wasn't for coarse-graining or partitioning. As will become clear below, I strongly agree with Robertson's take: to simply adopt an infinitely "fine-grained" partition would amount to not partitioning at all. It would render BSM dysfunctional. That being said, for the issues discussed here, it is irrelevant how the partition is justified, and whether this justification itself is justified or not. As long as we do partition phase space in the usual way, the issues discussed here arise, independently of the particular partition employed and its justification. I am indebted to an anonymous referee for the pointer to Robertson's account.

approach can rightly be criticized for partitioning phase space into macro regions. The Gibbsian approach is not under consideration here, so this issue shall not further defer the discussion. The point to be made is merely that others see a potential for criticism of the BSM framework as well, and locate it at the same juncture: the partition of phase space into macro regions with sharp boundaries.[13]

3.1.6 The Micro State Crossing Sharp Boundaries

Let's make DP apprehensible and reveal its consequences in a series of examples. Imagine a system of interest, e.g., a gas in a (sealed, insulated) room. Modelling it in the usual way, a partition of its available phase space including sharp boundaries between macro regions is introduced. These macro regions correspond to different macro states of the gas, such that, if the micro state of the gas is located within a certain macro region, the gas instantiates the macro state as specified by its corresponding macro values. Hence, the actual micro state of the gas determines its actual macro state. Let's say that, in its initial macro state M_0 at t_0, the gas is confined to a certain corner of the room. Let the system evolve until the gas has spread out and evenly fills the whole room. This state is called M_{eq} and marked with the time stamp t_{eq}. Clearly, the micro state representing the system must have evolved and crossed some boundaries. To simplify the picture, imagine that there are only two macro states: M_0 with the gas in the corner, and M_{eq} with the gas spread out. Accordingly, there are two macro regions, Γ_{M_0} and $\Gamma_{M_{eq}}$. Since the gas has changed

---

[13]The following exchange between Sheldon Goldstein and Daniel Sudarsky during the former's talk *Gibbs vs. Boltzmann Entropy* at the *Chimera of Entropy* summer school in Split in 2018 is also elucidating (my transcript):

Daniel Sudarsky: "Something that has bothered me all the time: consider that partition that you have [the usual partition with sharp boundaries]. Then consider a collection of points that are approaching the boundary from the two sides. I cannot imagine being able always to differentiate between them [… inaudible]".

Sheldon Goldstein: "You are right. So maybe you need to develop a better scheme which is more realistic, so you don't have sharp boundaries. […] You are quite right; in a better world than ours, somebody will have done that. Maybe somebody has some kind of fuzzy notion of macro state. But you don't want to introduce that here [in the context of a heuristic discussion of entropy and approach to equilibrium], which will just complicate things. But you are completely right, there isn't that realistic division between [macro regions]: 'Oh, these points over here look like equilibrium, go epsilon further, doesn't look like equilibrium anymore.' That, obviously, is unrealistic. But if you want to get an understanding of the phenomena you should make that idealisation." (https://www.youtube.com/watch?v=jU838SXBrv4, 00:52:30–00:53:40)

Of course, such a statement, made in the context of a presentation, must not be mistaken for the "official" position of the speaker. That is, it must not be assumed that Goldstein (or Sudarsky) would have made the same or a similar statement in a published paper. (Neither, that they wouldn't make such statements in a published paper. After all, (Goldstein et al., 2019) call a partition with sharp boundaries "artificial".) Nevertheless, this exchange can be regarded as a further hint to the effect that the way phase space is partitioned in the framework of BSM isn't unanimously viewed as entirely unproblematic.

its macro state from residing in the corner to filling the whole room, its micro state must have wandered from Γ_{M_0} to $\Gamma_{M_{eq}}$. Hence, it must have crossed the boundary between these macro regions. Zooming in on the phase space region where the micro state crosses the boundary between Γ_{M_0} and $\Gamma_{M_{eq}}$, there is a precise instant when this happens, namely when the micro state lies exactly on the boundary. Let's mark this instant $t_{crossing}$. Just before $t_{crossing}$, the micro state is in Γ_{M_0}, such that it is instantiating the macro state M_0, with the gas in the corner. Just after $t_{crossing}$, the micro state is in $\Gamma_{M_{eq}}$, such that it is instantiating the macro state M_{eq}, with the gas spread out. Hence, the continuous evolution of the micro state over just a tiny distance in phase space, such that it crosses the boundary, amounts to a rather dramatic discontinuity in the macro state of the gas in this model: at $t_{crossing}$, the volume jumps from a rather small value (the gas confined in the corner of the room) to a comparatively large value (the gas spread out over the entire room). Yet, judging from the viewpoint of thermodynamics, as well as from considerations about the speeds at which gas particles usually move—at finite speeds, certainly—one would expect a small, continuous evolution of the micro state only to amount to an accordingly small, and likewise continuous evolution of the macro state. After all, the continuous change of the micro state around $t_{crossing}$ might represent nothing but the displacement of one measly particle by a few microns. Such a tiny difference between micro states before and after $t_{crossing}$ can hardly account for the tremendous discontinuity between macro states M_0 and M_{eq}.

One might want to reply that this strange correlation between the microscopic and macroscopic evolution of the system is due to the overly simplified example used. Certainly, there are not only two macro states, and, accordingly, not only two macro regions with one boundary between them. The partition is way too coarse. But irrespective of how many macro regions are introduced in accordance with equally many macro states, the boundaries between macro regions are always sharp, and macro regions are of finite number, corresponding to a finite number of different macro states of the gas. So, no matter how fine-grained the partition is, as long as it contains finitely many macro regions of non-zero volume, corresponding to finitely many different macro states, the problem will prevail: a small, continuous change in the micro state of the gas, while crossing the boundary, will amount to a discontinuous change in its macro state. The issue can be put like this: as soon as boundaries are introduced when carving up phase space, i.e., as soon as a partition into macro regions is introduced, this entails the assumption that micro states "left" and "right" of these boundaries, i.e., micro states in different macro regions, instantiate different macro states. Otherwise, it would be nonsensical to introduce this partition. But once there are sharp boundaries, DP arises, no matter how fine- or coarse-grained the partition. Differentiating between fewer or more macro states results in coarser or finer graining of the partition, but partitions will always be grainy, as long as macro regions are of finite, non-zero volume.

3.1.7 Sharp Boundaries Being Shifted

The following example exploits the aforementioned arbitrariness of the partition. Imagine a partition of the available phase space of a system. Two neighbouring macro regions, Γ_{M_i} and Γ_{M_j}, correspond to different macro states M_i and M_j and are separated by a sharp boundary. Three micro states, x_1, x_2, x_3, are selected, such that x_1 lies somewhere in the middle of Γ_{M_i}, x_2 lies arbitrarily close to the boundary, but still in Γ_{M_i}, and x_3 resides equally close to the boundary, but opposite from x_2, in Γ_{M_j}. Figure 3.3 provides an illustration. A tiny change of the partition, i.e., a small shift of the boundary between Γ_{M_i} and Γ_{M_j}, should not amount to a model representing a completely different physical situation. Yet, this is what will happen. If the boundary indeed is drawn arbitrarily, it can be shifted, say, a wee bit towards Γ_{M_i}, such that the volume of Γ_{M_i} decreases ever so slightly, while the volume of Γ_{M_j} increases by the same amount. With x_2 lying arbitrarily close to the boundary, this shift results in both, x_3 and x_2, lying in Γ_{M_j} (Fig. 3.4).

This new, slightly different partition seems an appropriate correction to the previous one, at least at first sight: the micro states x_2 and x_3, always residing extremely close together, are now grouped into the same macro region Γ_{M_j}, thereby,

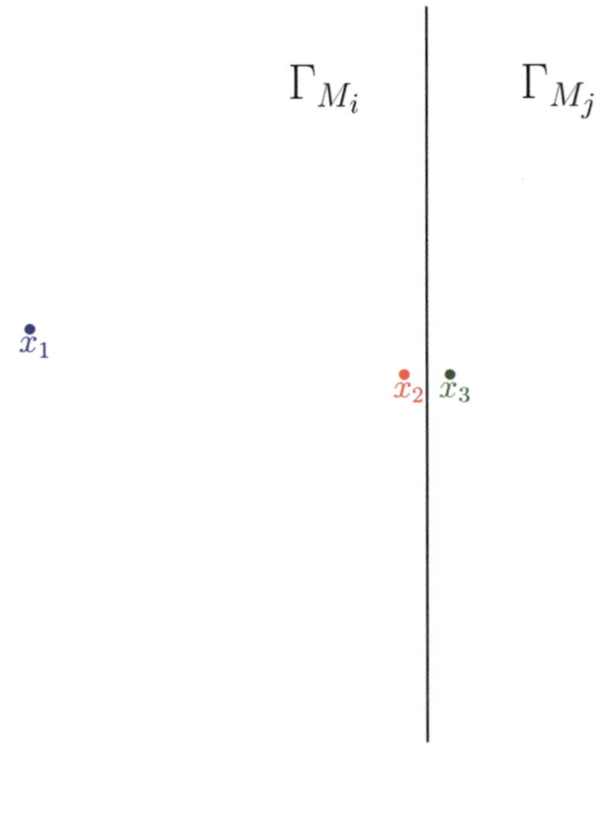

Fig. 3.3 Two neighbouring macro regions, separated by a sharp boundary

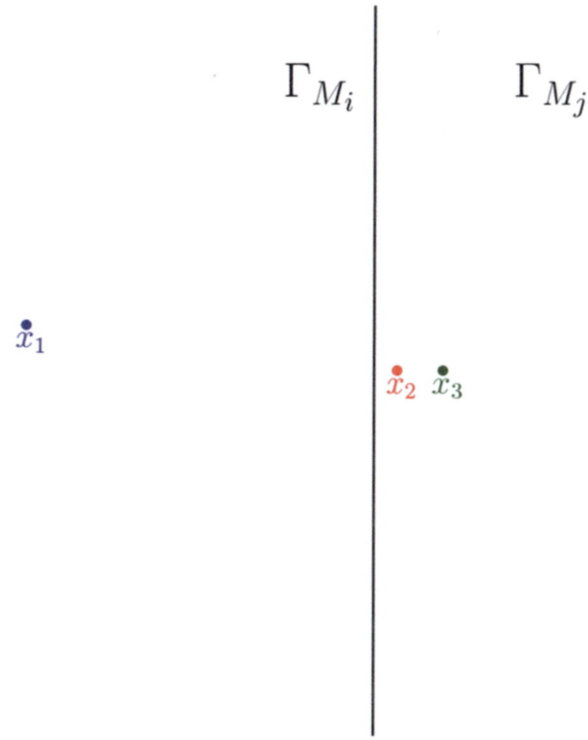

Fig. 3.4 The boundary shifted

according to the model, instantiating the same macro state M_j. The relatively remote micro state x_1 remains in its original region Γ_{M_i}. From a macroscopic point of view, x_2 and x_3 are now "macro-identified" in that they are taken to instantiate the same macro state. This macro state is different from the macro state instantiated by the fairly remote x_1.

But all is not well. A tiny change of the partition, this example shows, amounts to models representing completely different physical situations on the macro level while nothing on the micro level changes: before the boundary shift, x_1 and x_2 were regarded as instantiating the same macro state M_i, while x_3 was instantiating M_j. After the shift, x_2 and x_3 are instantiating the same macro state M_j, while x_1 is still instantiating M_i. So in case x_2 becomes actual, it makes a huge difference in the model whether the boundary has been shifted or not. At the same time, nothing in the target system's micro state has actually changed: x_2 specifies the exact same micro state as before the shift of the boundary. The disagreement between the two models, despite describing the same microscopic situation of the same target system, is evident. Merely changing the partition slightly while keeping the actualised micro state fixed results in different macro states being instantiated. This suggests that, on BSM, the macro state doesn't depend on the micro state *alone*, but also on the partition, which can be chosen rather freely. This is a troubling result.

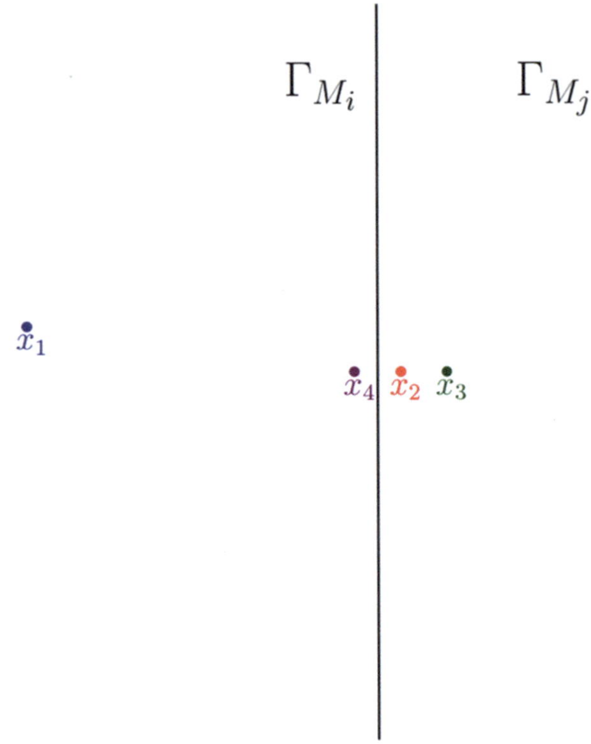

Fig. 3.5 Another micro state, selected arbitrarily close to the boundary

And there is more: x_1, x_2 and x_3 have been arbitrarily selected. They are just phase points, like any other. So why not select another micro state, call it x_4, also arbitrarily close to the (now shifted) boundary, opposite of x_2 (Fig. 3.5)?

The same problem as before reappears. So the "correction" of the partition turns out to be no correction after all. Shifting boundaries around, in effect, is just shifting around a discontinuity; it is shifting the problem to a slightly different location, thereby making the underlying problem more evident.

Sharp boundaries between macro regions in phase space, as these examples should have illustrated, are just as artificial as the border between the Netherlands and Belgium.[14] There is a line, drawn arbitrarily, but in its immediate vicinity, on both sides, things look quite the same.

[14] Incidentally, the border between Belgium and the Netherlands has shifted as recently as 2018. See Stone (2018).

3.1.8 The Case of Entropy

There is yet another, related issue arising from sharp boundaries that at least deserves mentioning. In the usual framework, the Boltzmann entropy function is generally defined as a function assigning an entropy value to macro states, depending on the volume of their respective macro regions: $S_B(M_i) := k_B \log\left[\mu\left(\Gamma_{M_i}\right)\right]$. The larger the volume, the higher the entropy. Thus, the largest region, i.e., the equilibrium region, is assigned the highest entropy. Accordingly, increase of Boltzmann entropy of a system[15] is described by the micro state moving to regions of larger volume (Goldstein et al., 2019, p. 19): "increase of Boltzmann entropy means that the phase point [x] moves to bigger and bigger macro sets [Γ_{M_i}]."

In a partition, different macro regions usually have different volumes, such that the micro state, upon traversing the boundaries between them, instantiates a macro state of different entropy. As a result, entropy changes in a jumpy fashion, like other macro values, as discussed above: upon the micro state moving from a smaller macro region to a larger one, entropy jumps from a lower to a higher value. The entropy function becomes discontinuous. By contrast, it stays at a constant value while the micro state wanders through a macro region, not crossing any boundaries. This behaviour, like the discontinuous behaviour of other macro values in BSM's models, might come as a surprise. Wouldn't one expect there to be a continuum of allowable entropy values, and that, while moving from one value to another, the continuum between them is instantiated, just like one would expect other macro values to evolve continuously? At least, it doesn't seem implausible to say that

[15] Boltzmann entropy of a system is defined as follows by Werndl and Frigg (2015b, p. 1225): "The Boltzmann entropy of a system at time t, $S_B(t)$, is the entropy of the system's macrostate at t: $S_B(t) := S_B\left(M_{x(t)}\right)$, where $x(t)$ is the system's microstate at t and $M_{x(t)}$ is the macrostate supervening on $x(t)$." See also the almost identical definition in Frigg and Werndl (2012, p. 102).

entropy and its change *ideally* would be described by a continuous function instead of a function that assigns the same entropy value during one finite time interval—the time interval during which the micro state traverses a macro region—and then another, different value during the finite time interval of traversal of the micro state through another macro region, with discontinuous jumps at the boundaries. But this is how entropy behaves in the models as they result from the standard framework. Arguably, this way of modelling entropy change is at odds with the continuous macro state evolution assumed for the thermodynamic target systems classical BSM seeks to describe (and reduce). As before, the origin of the issue can be traced back to the method of partitioning phase space into a finite set of disjoint macro regions with sharp boundaries and non-zero volume.

3.1.9 Modifying the Framework?

It has been argued that phase space partitioning with sharp boundaries poses a problem for modelling continuous macroscopic change in the usual framework of BSM. The examples above shall now be revisited while not assuming *exhaustive-cover* or *no-overlap*. These conditions, together with the requirement that macro regions are of non-zero volume, result in sharp boundaries. Let's see if, upon excising them, one at a time, the discontinuity problem vanishes.

3.1.9.1 Removing Exhaustive-Cover

First, *exhaustive-cover* shall be removed, thus allowing for models in which there are micro states that do not belong to a macro region: $\exists x \forall \Gamma_{M_i} \left(x \notin \Gamma_{M_i} \right)$. Regarding the partition, this means that there are regions of phase space that are neither a macro region, nor belong to one. Figure 3.6 provides an illustration. Via the correspondence between macro regions and macro states, phase points in those regions do not instantiate a macro state. Admittedly, a rather strange result. Under these conditions, it could well be that the micro state, while wandering through phase space, moves from a macro region to such a "non-region" and on to a macro region again, which would mean that, in between instantiating two macro states, for some time, it instantiates no macro state. But clearly, the constituents of the gas haven't lost their properties, all of a sudden. The phase points within such a non-region still describe the same microscopic features of the same system as the phase points in normal regions do: all the positions and momenta of all the system's constituents. Clearly, it is not the micro state of the gas that is responsible for this strangeness.

One might be tempted to resolve this issue by regarding these non-regions as transition regions between others. Recall the example of a gas in a room. The coarseness of this initial partition with only two macro regions can be reduced by introducing a "non-region" between these two that is taken to be a transition region.

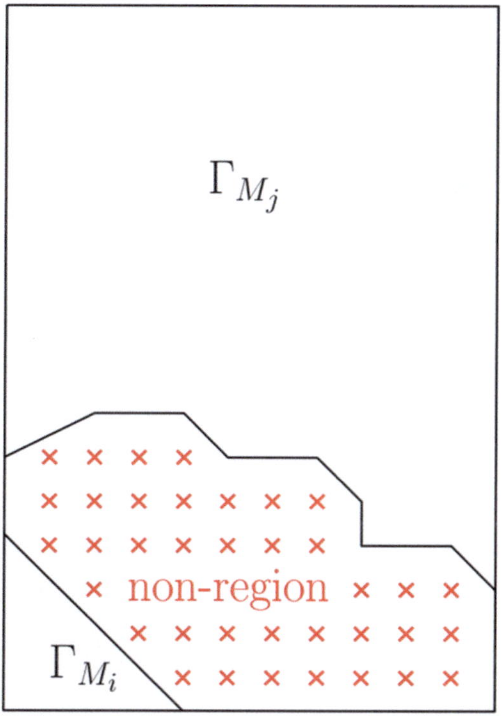

Fig. 3.6 Phase space partition with a non-region

Let's call the resulting macro regions Γ_{M_i}, Γ_{M_j} and $\Gamma_{M_{between}}$ (Fig. 3.7). That is all fine and well. The initial partition, after all, was too coarse. However, the so-called "non-region", then, is not a non-region, corresponding to no macro state at all. Instead, the points it contains instantiate a macro state $M_{between}$ between the initial state M_i and the final state M_j. Taking this route, so-called non-regions are just ordinary regions, such that there are no true non-regions in phase space. *Exhaustive-cover*, in effect, isn't removed.

Furthermore, apart from the strangeness of having true non-regions in phase space, even if one regards them as a perfectly normal feature of BSM's models, one would be facing the same discontinuity problem as before, and even more drastically. In transitions from ordinary regions to ordinary regions, a jump discontinuity between macro values appears. In transitions from ordinary regions to non-regions and vice versa, an essential discontinuity appears: one of the limits L^- or L^+ (see above) doesn't exist, since non-regions don't correspond to macro values. So there is not just a discontinuous jump between different macro values, but a jump from a macro value to an undefined value or vice versa, only exacerbating the issue.

As an interim result, it can be noted that it seems impossible to do away with the *exhaustive-cover* condition. Either, the resulting "non-regions" are true non-regions, making the discontinuity problem even worse, or they are ordinary regions after all. In any case, the original problem to be solved remains.

Fig. 3.7 The non-region taken to be a transition region

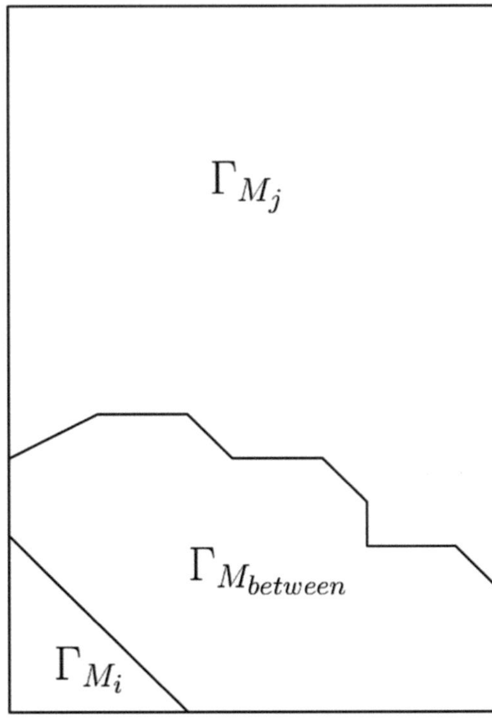

3.1.9.2 Removing No-Overlap

So let's try keeping *exhaustive-cover* and excising *no-overlap* instead. This allows for micro states that belong to more than one macro region: $\exists x \exists \Gamma_{M_i} \exists \Gamma_{M_j} \left(x \in \Gamma_{M_i} \wedge x \in \Gamma_{M_j} \right); i \neq j$. Accordingly, there can be overlap regions in phase space (Fig. 3.8).

A micro state within such an overlap region $\Gamma_{M_i} \cap \Gamma_{M_j}$ is in Γ_{M_i} as well as Γ_{M_j}, and therefore instantiates both, M_i and M_j. As long as these have the same macro values, all is fine. However, they can't. As per usual, since macro states are specified via macro values, and correspond to macro regions, macro states being specified by the same macro values are identical macro states, i.e., there is only one macro state, corresponding to one macro region. In other words: macro regions corresponding to identical macro states form a union into one macro region. So, whenever one wants to speak of different macro regions, they must correspond to different macro states. And that, in turn, means that an overlap region of two different macro regions corresponds to two different macro states. Accordingly, every micro state in an overlap region instantiates two different macro states at the same time, which is an inconsistency.

To provide an extreme example of this equally strange result, recall again the simple gas example. Imagine that there is some overlap $\Gamma_{overlap} = \Gamma_{M_0} \cap \Gamma_{M_{eq}}$. If the

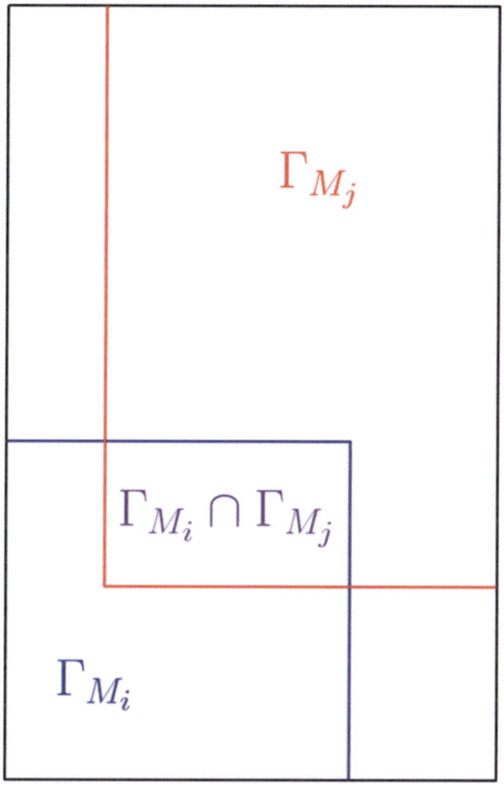

Fig. 3.8 Phase space partition with overlapping regions

micro state is in $\Gamma_{overlap}$, it is in Γ_{M_0} and $\Gamma_{M_{eq}}$. Accordingly, the gas is concentrated in the corner (M_0) as well as evenly spread out over the whole room (M_{eq})—an inconsistency even worse than DP. Hence, one should abstain from excising *no-overlap* as well.

3.1.9.3 A Different Solution?

Removing either *exhaustive-cover* or *no-overlap* from BSM's framework results in severe problems. It thus seems reasonable to keep them. Yet, at the same time, leaving everything at status quo provides no solution to the discontinuity problem as it arises from sharp boundaries between a finite number of macro regions of non-zero volume. Consequently, one may ask whether a framework in which a set of phase points, i.e., a macro region, corresponds to *one and the same* macro state is a good one. Tweaking the framework's conditions didn't result in the desired solution. Maybe, then, it is time to consider adopting a different one.

3.1.10 The Discontinuity Problem and the Foundational Project

Let's finally dwell on the consequences DP has for the foundational project of reducing thermodynamics to BSM. As shown above, T_{TD} and T_{BSM} behave differently. This discrepancy lies at the heart of DP. But why is this a problem for the foundational project? Presumably the simplest and, as Sklar (1993, 340 f.) writes, "one of the primary ways in which our theories of the world are unified" is "reduction by identification". "A general methodological principle seems to be that we ought to identify, as opposed to positing a correlation, whenever we can. That is, we should identify whenever the assertion of such an identification is not blocked by some feature of the situation, typically by the reduced entity having some feature genuinely not attributable to the reducing". When trying to reduce thermodynamics to BSM, one encounters such a blocking feature. The reduced theory (thermodynamics) has a feature, namely continuous change of macro values, that is not attributable to the reducing theory (BSM), which can handle only discontinuous change of macro values.

As Sklar points out, the reduction of thermodynamics to statistical mechanics *does* involve a genuine identification, insofar as it takes part in the broader reductive program (ibid., 341): "This is the reduction of the theory of macroscopic matter to its micro-constituents by the identification of the macroscopic entities as structured out of microscopic entities." BSM successfully does that. It takes macroscopic systems, e.g. gases, to be structured out of microscopic entities (ibid., 348): "the thermodynamics of gases can only be reduced to statistical mechanics after we have already identified gases as collections of interacting molecules and reduced our gas theory to a theory of molecules and their interaction." But he is also eager to point out that this is not the whole story—that the matter is more intricate (ibid., 341): "whether the reduction also involves, in any simple way, the identification of the thermal features of things (especially temperature and entropy) with features of things characterized at the reducing level in non-thermodynamic terms is a subtler matter." One could, for example, put into question whether the thermodynamic concept of temperature, which is reasonably attributed to macroscopic systems, can really be identified with the "kinematic-dynamic" (Sklar, 362) concept of average kinetic energy of the microscopic constituents in terms of their velocities and masses. At the very least, such a conceptual identification is certainly not trivial. But one doesn't even have to go this far in order to encounter a problem. The thermodynamic macroscopic temperature value of a target system (T_{TD}) can change continuously. The macroscopic temperature value of the BSM model (T_{BSM}) describing the target system can not. Already these closely related concepts, both denoting temperature on the same, macroscopic level of description, are not identified trivially. It is easy to see this by asking: does T_{TD} supervene on T_{BSM}? The answer is "no". In order to have T_{TD} supervene on T_{BSM}, the latter must change whenever the former changes, which is not the case, due to DP. Supervenience is certainly necessary for identification, and a fortiori for reduction qua identification. So, since T_{TD} doesn't supervene on T_{BSM}, reduction of the former to the latter qua identification is blocked. Adopting the terminology from Wimsatt (2006), one could say that the intra-level reduction from thermodynamics to BSM fails for the macro-quantity temperature.

This is quite unlike the reduction (qua identification) of the concept of a gas (pre kinetic gas theory) to the concept of a gas (post kinetic gas theory) in terms of a collection of interacting, particular constituents. The thermodynamically described, macroscopic gas clearly supervenes on its constituents: macroscopic, thermodynamically described change of the gas is always accompanied by microscopic, kinematic-dynamic change, as reflected by the fact that the micro state evolves through phase space. That is, whenever T_{TD} changes, the micro state changes as well. But not so the BSM temperature macro value T_{BSM}. In BSM's models, T_{BSM} remains fixed for a finite interval of time (during the micro state's evolution within a macro region) while, during the same time interval, the thermodynamic temperature value T_{TD} changes. So T_{BSM} doesn't always change whenever T_{TD} changes. So T_{TD} does not supervene on T_{BSM}. Instead, the supervenience relation holds the other way round: whenever T_{BSM} changes, T_{TD} changes as well. However, this is the wrong way for the foundational project. For a reduction of thermodynamics to BSM to be possible, T_{TD} must supervene on T_{BSM}. In effect, the foundational project is partially successful, and partially not. It is successful in that thermodynamically described systems can be reduced to systems of interacting constituents qua identification. That is, *inter*-level reduction succeeds. It is unsuccessful in that macro properties as described by thermodynamics cannot be reduced to macro properties as described by the models of BSM, because, as the exposition of DP establishes, the former do not supervene on the latter. Hence, *intra*-level reduction fails.

3.1.11 Does T_{TD} Supervene on T_{BSM}?

Above, I have argued that T_{TD} cannot be reduced to T_{BSM} because the former does not supervene on the latter. In this section, I want to make this failure of supervenience, and a fortiori of reduction, explicit in a more technical way.[16]

For the sake of the following argument, we restrain ourselves to the classical, kinetic gas theory.[17] Here, thermodynamic temperature T_{TD} is conceptualized as a

[16] I am indebted to an anonymous referee who pressed me on spelling out this argument in greater technical detail.

[17] If one should be concerned about the assumption of kinetic gas theory and insist on construing gases as continua for thermodynamics, the same argument can be construed without referral to individual gas particles by imagining the arbitrarily small increase of the temperature of such a continuous gas by an arbitrarily small amount of energy transferred into it from an external system, as heat or work exerted on it. The main purpose of construing the argument based on kinetic gas theory not only for the BSM framework but also for thermodynamics is to emphasize that the problem of reduction discussed here is nothing to do with the underlying physical ontologies assumed for the respective theories. The argument gains clarity by assuming the same physical

measure of the average kinetic energy $\overline{E_{kin}}$ of the constituents of a system, e.g., of the particles of an ideal, monatomic gas. The kinetic energy of a single particle is given as

$$E_{kin} = \frac{1}{2}mv^2$$

with m denoting the mass of the particle and v its velocity. Since in an ideal, monatomic gas m is the same for all particles, the average kinetic energy of all particles of the gas is given as

$$\overline{E_{kin}} = \frac{1}{2}m\overline{v^2}$$

where the bars denote averages, i.e., $\overline{v^2}$ is the average squared velocity of the particles and, accordingly, $\overline{E_{kin}}$ their average kinetic energy.

With the above conceptualization of thermodynamic temperature as a measure of average kinetic energy, T_{TD} is directly proportional to $\overline{E_{kin}}$:

$$T_{TD} \propto \overline{E_{kin}}$$

Via the ideal gas law, with k_B denoting the Boltzmann constant, this proportionality is expressed as

$$\overline{E_{kin}} = \frac{1}{2}m\overline{v^2} = \frac{3}{2}k_B T_{TD}$$

i.e.

$$T_{TD} = \frac{m\overline{v^2}}{3k_B}$$

$\frac{m}{3k_B}$ is a positive, real-valued constant. For simplicity, let's call it κ , such that

$$T_{TD} = \kappa\overline{v^2}$$

ontologies for both, but it goes through even without this assumption, since the intertheoretic reduction we are after is an issue of how the theoretical properties of the theories—the values of the variables with which they represent the phenomenon of macroscopic temperature and its change—relate to each other: whether they instantiate the appropriate supervenience relation or not.

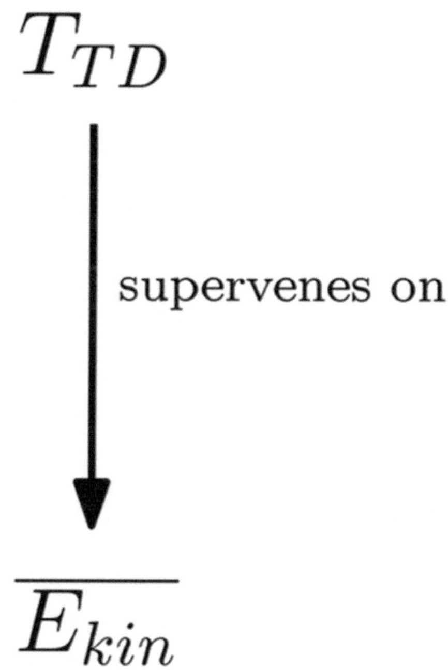

Fig. 3.9 Thermodynamic temperature supervenes on average kinetic energy

This shows the direct correspondence between thermodynamic temperature and average particle velocity in the case of ideal, monatomic gases.

Imagine now that the thermodynamic temperature increases by some amount ΔT_{TD}. With the above relation $T_{TD} = \kappa \overline{v^2}$, where κ is constant, some corresponding increase of the average particle velocity squared must account for it. And with $\overline{E_{kin}} = \frac{1}{2} m\overline{v^2}$, this corresponds to an increase in average kinetic energy. Thus, whenever T_{TD} changes, average kinetic energy changes as well. The former supervenes on the latter (Fig. 3.9).

We can even show that this supervenience relation holds in the opposite direction, too. Imagine that the velocity of one particle is increased by some amount Δv while everything else remains unchanged. Already intuitively, it is clear that this increase in velocity of one particle amounts to a proportional increase of T_{TD}:

$$\Delta v \propto \Delta T_{TD}$$

The average velocity (i.e., the arithmetic mean)[18] of all particles of the system at time t_1 is the sum of the velocities of the individual particles over the number N of particles:

[18] Instead of the arithmetic mean one could use the root mean square (rms). However, the difference is irrelevant for the present argument; taking the rms would merely complicate calculations without any gain in clarity. See, e.g., Giancoli (2010, p. 630f.).

$$\overline{v_{t_1}} = \sum_{i=1}^{N} \frac{v_i}{N}$$

If at t_2 the velocity of one particle is increased by Δv, this increases to

$$\overline{v_{t_2}} = \left(\sum_{i=1}^{N} \frac{v_i}{N} \right) + \frac{\Delta v}{N}$$

Thus, the average velocity $\overline{v_{t_2}}$ at time t_2 is larger than the average velocity $\overline{v_{t_1}}$ at t_1 by $\frac{\Delta v}{N}$:

$$\overline{v_{t_2}} = \overline{v_{t_1}} + \frac{\Delta v}{N}$$

The average squared velocity of all particles of the system at time t_1 is the sum of the squared velocities of the individual particles at t_1 over the number N of particles:

$$\overline{v_{t_1}^2} = \sum_{i=1}^{N} \frac{v_i^2}{N}$$

If the velocity of one particle—particle 1, say—is increased by Δv_1, this increases to

$$\overline{v_{t_2}^2} = \left(\sum_{i=1}^{N} \frac{v_i^2}{N} \right) + \frac{2v_1 \Delta v_1 + \Delta v_1^2}{N} = \overline{v_{t_1}^2} + \frac{2v_1 \Delta v_1 + \Delta v_1^2}{N}$$

When plugging this into

$$T_{TD} = \kappa \overline{v^2}$$

we see that T_{TD2} must be larger than T_{TD1}, by the amount $\kappa \left(\frac{2v_1 \Delta v_1 + \Delta v_1^2}{N} \right)$:

$$\begin{aligned} T_{TD2} &= \kappa \overline{v_{t_2}^2} = \kappa \left(\overline{v_{t_1}^2} + \frac{2v_1 \Delta v_1 + \Delta v_1^2}{N} \right) = \kappa \overline{v_{t_1}^2} + \kappa \left(\frac{2v_1 \Delta v_1 + \Delta v_1^2}{N} \right) \\ &= T_{TD1} + \kappa \left(\frac{2v_1 \Delta v_1 + \Delta v_1^2}{N} \right) \end{aligned}$$

As long as v_1 and Δv_1 share the same sign ($v_1 \Delta v_1 > 0$) and Δv_1 is finite, i.e., as long as Δv_1 is an actual increase of velocity in the same direction as v_1 (or in any direction, if v_1 is 0), it is always the case that

$$\kappa\left(\frac{2v_1\Delta v_1 + \Delta v_1^2}{N}\right) > 0$$

and thus that

$$T_{TD2} > T_{TD1}$$

This shows that, in thermodynamics, an increase in the velocity of one constituent particle of a system, with the other parameters held fixed, amounts to an increase in temperature of that system. Importantly, this Δv can be arbitrarily small; as long as it isn't zero, thermodynamic temperature increases. Thus, whenever average kinetic energy changes, T_{TD} changes as well. The former supervenes on the latter (Fig. 3.10).

That T_{TD} and average kinetic energy supervene on each other doesn't come as a surprise. This symmetric supervenience is a necessary condition for the background reduction qua identification already mentioned above. When temperature is conceptualized as a measure of average kinetic energy, it is only to be expected that the two supervene on each other.

How does this situation present itself in the BSM framework? There, the micro state of a system consisting of N constituents is given as a point x in a $6N$-dimensional phase space Γ, where the three position degrees of freedom and the three momentum degrees of freedom of each of the N particles are represented by one

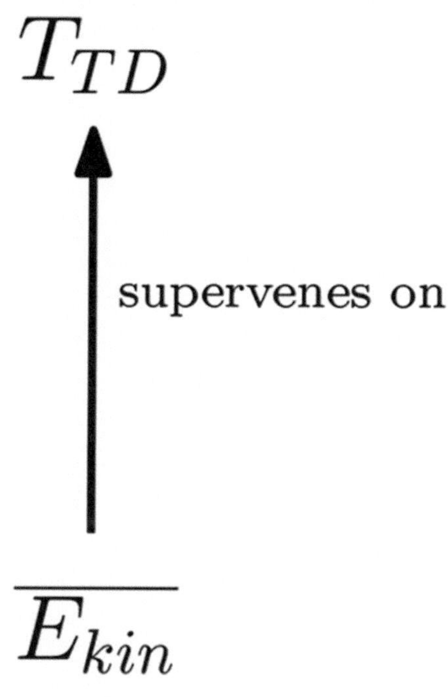

Fig. 3.10 Average kinetic energy supervenes on thermodynamic temperature

dimension each. In other words, the micro state is represented by an ordered $6N$-tuple $(q_1, q_2, \ldots, q_{3N}, p_1, p_2, \ldots, p_{3N})$, where the q_i each represent the position of one particle in one spatial dimension, and the p_i each represent the momentum of one particle in one spatial direction. Since momentum is defined as the product of mass and velocity, $p = mv$, it can be increased by increasing velocity: $p + \Delta p = m(v + \Delta v)$. With $p = mv$, the increase of momentum is

$$\Delta p = m\Delta v$$

As before, suppose the velocity of one particle—particle 1—is increased by Δv in a certain direction, e.g., the second spatial direction, such that its momentum is increased by Δp in that direction as well. That is, the micro state changes from

$$x_1(q_1, q_2, \ldots, q_{3N}, p_1, p_2, \ldots, p_{3N}) \text{ to } x_2(q_1, q_2, \ldots, q_{3N}, p_1, p_2 + \Delta p_2, \ldots, p_{3N}).$$

Choose Δv such that the resulting Δp_2 is so large that x_2 resides in a different macro region than x_1, i.e., that x_2 instantiates a different macro state than x_1 does. Then, $T_{BSM2}(x_2) \neq T_{BSM1}(x_1)$. In fact, whenever $T_{BSM2}(x_2) \neq T_{BSM1}(x_1)$, $x_2 \neq x_1$, because T_{BSM1} and T_{BSM2} correspond to different macro regions, and macro regions don't overlap, such that x_1 and x_2 cannot be identical (see above). Since the only difference between x_1 and x_2 is a difference in the velocity of one particle, x_1 and x_2 differ in average kinetic energy. Thus, whenever T_{BSM} changes, average kinetic energy changes as well. The former supervenes on the latter (Fig. 3.11).

But now recall from above that Δv can be chosen arbitrarily small. If Δv can be arbitrarily small, then so can be Δp. Make it so small that the micro state x, after

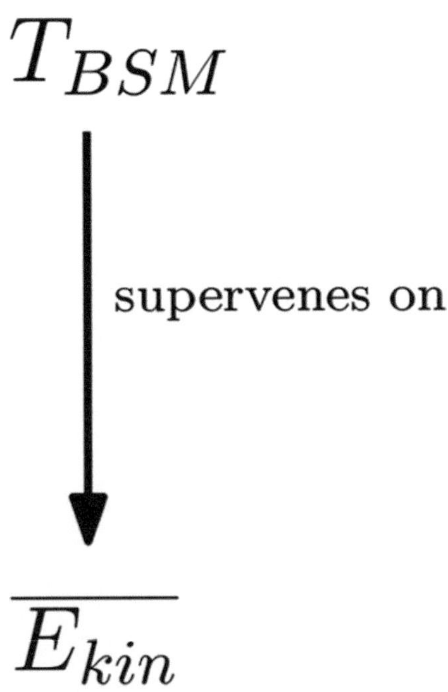

Fig. 3.11 BSM temperature supervenes on average kinetic energy

increasing the velocity of particle 1 by Δv in the second spatial direction, remains in the same macro region in the BSM picture. I.e., x_1 and x_2 are both elements of the same macro region: $x_1, x_2 \in \Gamma_{M_i}$. But since all micro states in Γ_{M_i} are assigned the same temperature macro value T_{BSMi} via the one-to-one correspondence between macro regions and sets of macro values, x_1 and x_2, according to the BSM picture, instantiate the same temperature: $T_{BSM1}(x_1) = T_{BSM2}(x_2)$. Since the difference between x_1 and x_2 is a difference in the velocity of one particle, x_1 and x_2 differ in average kinetic energy. At the same time, T_{BSM1} and T_{BSM2} don't differ. Thus, average kinetic energy doesn't supervene on T_{BSM} (Fig. 3.12).

We have seen that, if Δv is sufficiently small, $T_{BSM1}(x_1) = T_{BSM2}(x_2)$. At the same time, according to the thermodynamic picture, for the same Δv, $T_{TD2} > T_{TD1}$. Even though in both cases, the velocity of the same particle was increased in the same direction by the same amount Δv, thermodynamic temperature T_{TD} did change while BSM-temperature T_{BSM} did not. Thus, there are cases in which T_{TD} changes while T_{BSM} does not change. And since supervenience of *A* on *B* requires that, whenever *A* changes, *B* changes as well, T_{TD} does not supervene on T_{BSM} (Fig. 3.13).

Fig. 3.12 Average kinetic energy does not supervene on BSM temperature

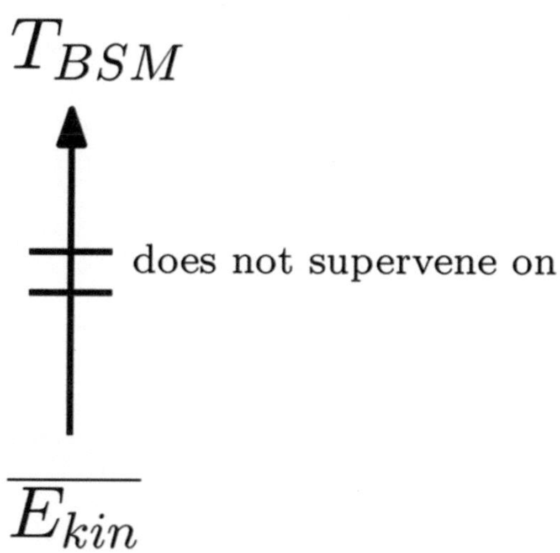

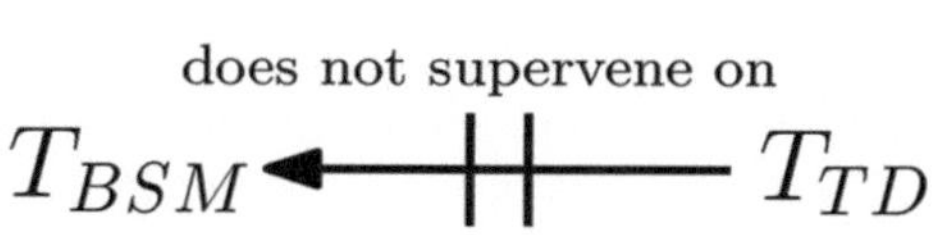

Fig. 3.13 Thermodynamic temperature does not supervene on BSM temperature

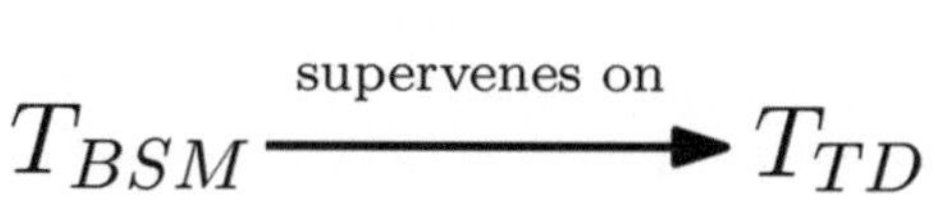

Fig. 3.14 BSM temperature supervenes on thermodynamic temperature

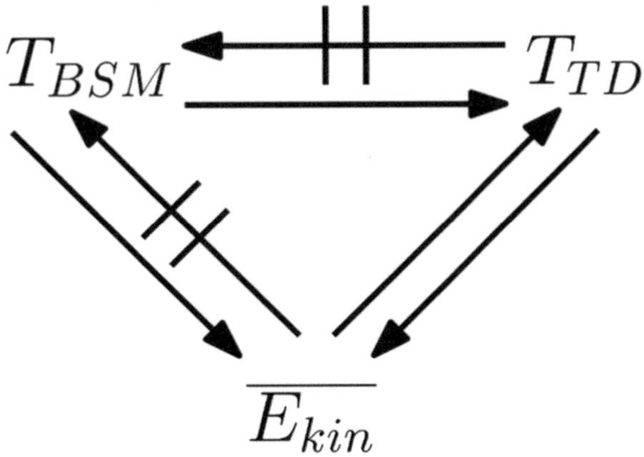

Fig. 3.15 Figures 3.9 to 3.14 put together

Conversely, T_{BSM} does supervene on T_{TD} (Fig. 3.14): whenever Δv is chosen sufficiently large that $T_{BSM1} \neq T_{BSM2}$, it is also sufficiently large that $T_{TD1} \neq T_{TD2}$, because every finite Δv suffices for that (see above).

Let's put the pieces together. We have seen that (a) T_{BSM} supervenes on average kinetic energy, but not vice versa; (b) T_{TD} supervenes on average kinetic energy and vice versa; (c) T_{BSM} supervenes on T_{TD}, but not vice versa (Fig. 3.15).

That T_{TD} does not supervene on T_{BSM} (but vice versa) is the most important result. It has been established by the independent arguments above, but it can also be seen when tracing the supervenience relations via average kinetic energy: T_{BSM} does supervene on average kinetic energy, and the latter supervenes on T_{TD}. A fortiori, since supervenience is transitive, T_{BSM} does supervene on T_{TD}. Conversely, while T_{TD} does supervene on average kinetic energy, the latter does not supervene on T_{BSM}. Hence, T_{TD} does not supervene on T_{BSM}.

One might be tempted to respond that the failure of supervenience of T_{TD} on T_{BSM} is all due to the specific way the partition was set up, and that partitioning into ever smaller macro regions would circumvent the issue. The first half of this response would be correct: indeed, the issue arises because of the partition. However, the second half of the statement is not correct: partitioning Γ into ever smaller macro regions is of no avail, since Δv can be chosen arbitrarily small. Without modifying the BSM framework substantially, the issue could be avoided only if one chooses to make the partition so small that, ultimately, every macro region contains only one micro state; that, however, amounts to the same as not partitioning at all. It is the same as directly assigning sets of macro values to individual phase points, which would be a nice thing to do, if we were able to do it. However, one of the main perks of SM is to circumvent the fact that we can't. On top of that, we'd be giving up advantages of the BSM framework, e.g., the possibility to define Boltzmann entropy as a measure of the size of macro regions: $S_B\left(M_i\right) := k_B \log\left[\mu\left(\Gamma_{M_i}\right)\right]$. With an infinitely fine-grained partition, not only would all macro states be assigned the same entropy, but that entropy value would be undefined—a situation that can hardly be satisfactory.

3.1.12 Does Thermodynamics Reduce to BSM?

We have just seen that T_{TD} does not supervene on T_{BSM}. As van Riel and Van Gulick (2019, 4.5.3) write, supervenience is a *conditio sine qua non* for reduction: "[Supervenience] surely is a necessary condition for reduction." McLaughlin and Bennett (2018, 3.3, my emphasis) even maintain that this conviction is held unanimously: "*Everyone* agrees that reduction requires supervenience." And they go on: "[O]n any reasonable view of reduction, if some set of A-properties reduces to a set of B-properties, there cannot be an A-difference without a B-difference."[19] In the case discussed here, which is a case of intertheoretic reduction, the sets of A- and B-properties mentioned by McLaughlin & Bennett are properties of the respective

---

[19] An anonymous referee has pointed out that supervenience might not be a *conditio sine qua non* for *all* kinds of reduction. I concur; indeed, it is possible to challenge the universality expressed in these quotations. However, for the kind of reduction envisaged by the foundational project, I think that supervenience *is* necessary. To flesh this out with two examples, suppose that supervenience isn't necessary for reduction. Then, reduction is possible without supervenience. That is, it is then possible to reduce some A property to some B property without B changing whenever A changes. First, it is then possible to reduce the temperature of an ideal, monatomic gas to its micro state without the latter changing whenever the former changes. That, however, entails that the gas temperature can change without the micro state undergoing any change whatsoever. At least in my view, this is inconceivable: how would a physical macro property of a physical object change without the physical micro state of the *very same* object changing as well? Such a position would undermine the very basic assumption of physical composition: that macroscopic objects such as gases are made up of smaller stuff, atoms in this example, and that their properties and behaviour arise from the properties and behaviour of the smaller stuff. And because of that, it would undermine the background "reduction of the theory of macroscopic matter to its micro-constituents by the identification of the macroscopic entities as structured out of microscopic entities." (Sklar, 1993, p. 341.) As Sklar points out, without this background reduction, the reduction of thermodynamics to statistical mechanics is impossible.

Similarly, in the case of the temperature macro property: if supervenience wasn't necessary for reduction, one could reduce T_{TD} to T_{BSM} without the latter changing whenever the former changes. Surely, that would alleviate the problem discussed here, but it is questionable whether this approach is reasonable and in the spirit of the foundational project. How would a macro variable (T_{TD}), describing a physical macro property of a physical object, change, without another macro variable (T_{BSM}), intended to describe the *very same* physical macro property of the *very same* physical object, changing as well? As long as one engages in the foundational project, i.e., as long as one aims to fully reduce TD to BSM, one must ensure that BSM's models adequately track those of TD. And in order to ensure that, in turn, T_{BSM} must change whenever T_{TD} changes. And that is nothing other than to say that T_{TD} must supervene on T_{BSM}. In short: for the foundational project, I see no way around the supervenience condition.

That being said, it might be worthwhile to try to come up with a kind of reductionist account that (i) is in the spirit of the foundational project and the reduction it envisages; (ii) does not do away with very basic assumptions such as physical constitution; but (iii) does not require supervenience. It can already be said that this kind of reduction, then, must adhere to some constraints. For example, it must not simply be a limiting reduction (see Palacios, 2019 for an enlightening demarcation of limiting reduction from other kinds of reduction) where the size of macro regions goes to zero, for the reasons extensively discussed above. This, however, is a separate—and intricate—project that lies outside the scope of this chapter.

theoretical frameworks, i.e., thermodynamics and BSM. More precisely, they each are the theoretical elements that represent the temperature of a target system in their respective theoretical frameworks, the values of the temperature variables T_{TD} and T_{BSM}, respectively. For thermodynamics to reduce to BSM, according to the statements just quoted, the value of T_{BSM} should change whenever the value of T_{TD} changes. But as we have seen, this is not always the case: T_{TD} does not supervene on T_{BSM}. Thus, thermodynamics does not reduce to BSM.

With supervenience being a *conditio sine qua non* for reduction, any kind of reduction of thermodynamics on BSM is blocked. But one might want to elaborate a bit. I am willing to concede that BSM partially—but *only* partially!—reduces thermodynamics, if only in a way that is not exclusive to BSM: the particular way in which BSM reduces thermodynamics is available even without BSM. Thus, this reduction of thermodynamics isn't BSM's merit.

Van Riel and Van Gulick (2019, 2.1) distinguish between two main strands of reduction, *diachronic* and *synchronic* reduction. The former construes reduction, amongst being "a temporal affair", as a process of "replacement of one theory by another theory, such that one theory (the reducing one) becomes the successor of the reduced theory".

However, for one theory to become the successor of another in this sense, it is not sufficient that it comes after its predecessor, in the temporal sense. This is just one aspect of the successor relation. Usually, this relation also comprises, as Kemeny and Oppenheim (1956, p. 7) put it, "the replacement of an accepted theory (or body of theories) by a new theory (or body of theories) which is in some sense superior to it. Reduction is an improvement in this sense." Van Riel and Van Gulick (2019, 2.1, my emphasis) elaborate on this: "[Diachronic reduction] is also often described as a sort of theory-improvement or scientific progress: particularly one in which *the laws of the prior theory apply to a proper subset of the cases covered by the laws of the succeeding theory*".

An example that gets cited for this kind of reduction is the alleged reduction of Newtonian dynamics of massive bodies to special relativity: the latter came not only after the former, it also improved on it, in the sense that it covered a wider range of phenomena, while still covering non-relativistic cases, i.e., cases in the low velocity domain, where $v \quad c$. Thus, Newtonian laws apply to a proper subset of cases of special relativity, and consequently, the second requirement for diachronic reduction, apart from temporal succession, is fulfilled, at least in this "narrow demonstration" (Fletcher, 2019).

Whether this really suffices to make the case for a successful reduction is a matter of ongoing debate, see Fletcher (2019). But at least the supervenience requirement is fulfilled for the crucial variables. For example, whenever Newtonian momentum $p_{Newt} = mv$ changes, relativistic momentum $p_{Relat} = \dfrac{mv}{\sqrt{1-(v/c)^2}}$ changes as well. Thus, p_{Newt} supervenes on p_{Relat}. At least, there is no issue of failed supervenience standing in the way of this (alleged) example of diachronic reduction.

The reduction of the concept of a gas (pre kinetic gas theory) to the concept of a gas (post kinetic gas theory) is an example of successful diachronic reduction, see above. What's more, it is also an example of successful intertheoretic reduction in the Nagelian sense.

The Nagelian model describes intertheoretic reduction as an explanation relation between two theories (see van Riel & Van Gulick, 2019, 2.2.1). Insofar explanation is demanded of successful, intertheoretic reduction, one can state that indeed, BSM explains, to some extent, thermodynamics, and, in this restricted sense, reduces it successfully. However, the qualification "to some extent" is important here, because it actually is the "background reduction" of "our gas theory to a theory of molecules and their interaction" (Sklar, 1993, p. 348) that does all the work, qua some identity assumptions: in the exemplary case of a gas, the spatial distribution of its constituent particles is identical to the (macroscopic) volume V of the gas and thus explains it; the force per area the particles exert on the walls of a container is identical to the (macroscopic) pressure P of the gas in that container and thus explains it; the average kinetic energy of the particles is identical to the (macroscopic) temperature T of the gas and thus explains it; the transfer of kinetic energy from particles of one system to those of another, and with it the redistribution of particle velocities, is identical to conductive heat flow from a hotter to a colder body after they are brought into contact, and thus explains it; etc. The background assumptions relevant for these explanations are part and parcel of kinetic gas theory. But kinetic gas theory, while it also features in BSM, can be had without BSM. And the other crucial part of BSM, its partitioning of the available phase space of a system, doesn't contribute to this explanation. In fact, it introduces difficulties: even while gases are identified with collections of particles in BSM, and macroscopic temperature T is thought of as average kinetic energy of these particles, BSM-temperature T_{BSM}, qua BSM's association of macro values with macro regions, merely approximates T, unless the partition is so fine-grained that it becomes dysfunctional and pointless. So, while a background assumption of BSM does all the reductive work, and a fortiori, BSM in this restricted sense does reduce thermodynamics, the complete BSM framework is not an integral part of this reduction, and T_{BSM} may not be identified with temperature in the sense of average kinetic energy, although it supervenes on it. Thus, BSM, as opposed to the kinetic gas theory, does not reduce temperature qua identification. Hence the caveat.

Most importantly, however, this background reduction is not the reduction we are after. Our question was whether T_{TD} reduces to T_{BSM}. And while both, T_{TD} and T_{BSM}, supervene on average kinetic energy, via their common background assumption of kinetic gas theory, T_{TD} does not supervene on T_{BSM}, and hence cannot be reduced to it. In other words, while prima facie one might be led to believe that the notion of temperature is the same in thermodynamics and BSM, it is, in fact, not.

And there is more. On the Nagelian account of reduction (and successors of it, e.g. the Generalised Nagel-Schaffner Model of Reduction, see Dizadji-Bahmani et al., 2010), explanation of the reduced theory by the reducing theory stems from derivability: the laws of the (corrected) reduced theory must be derivable from the laws of the reducing theory plus "some auxiliary assumptions" and bridge laws in order to explain and thus reduce it. The law of the reduced theory we are concerned with here is the (corrected) law governing the time evolution of T_{TD}; the law of the reducing theory is the law governing the time evolution of T_{BSM}. So, the question arises: can the (corrected) law governing the time evolution of T_{TD} be derived from the law governing the time evolution of T_{BSM} (plus bridge law and auxiliary assumptions)? At least for a "corrected" version of thermodynamics, it would be unnecessary to recover the exact, uncorrected thermodynamic treatment from BSM. Unfortunately, this derivation is not possible. On the one hand, the only candidate auxiliary assumption that would render it possible is an infinitely fine-grained partition. But, as I have argued, this route is not available, in particular because it leaves BSM dysfunctional and would amount to the same as not partitioning at all, such that we are left with the description of the micro state and its time evolution, disregarding macroscopic regularities (see fn. 12 above). But the law governing the time evolution of T_{TD} is not derivable from the law governing the time evolution of the micro state alone, without a proper partition, not even only in approximation. If it were, we wouldn't need BSM in the first place. On the other hand, when applying the usual partition, at least we have a candidate bridge law, something that looks like a bridge law, but unfortunately, isn't one. Dizadji-Bahmani et al. even provide the reason why there is no bridge law establishing the relevant correlation between T_{TD} and T_{BSM}. They write (399. I have replaced their placeholders for theoretical terms, t_P and t_F, by the ones that are relevant here, T_{TD} and T_{BSM}): "A bridge law is a statement to the effect that (1) T_{TD} applies if, and only if, T_{BSM} applies, and (2) $\tau_{T_{TD}} = f\left(\tau_{T_{BSM}}\right)$", where $\tau_{T_{TD}}$ is the value of T_{TD} and $\tau_{T_{BSM}}$ the value of T_{BSM}. $f\left(\tau_{T_{BSM}}\right)$ is a function that transforms $\tau_{T_{BSM}}$ into $\tau_{T_{TD}}$, i.e., the BSM temperature value into the thermodynamic temperature value. However, in the case we are discussing here, f is not even a function, because uncountably many $\tau_{T_{TD}}$ correspond to every $\tau_{T_{BSM}}$. Hence $f\left(\tau_{T_{BSM}}\right)$ does not map each element of $\tau_{T_{BSM}}$ (its domain) to exactly one element of $\tau_{T_{TD}}$ (its codomain). The second requirement for the bridge law isn't met, and thus, there is no bridge law connecting T_{TD} and T_{BSM} in the necessary way, even if the two are coextensive. Note that this has nothing to do with multiple realisability, which states that a property of the reduced theory may correspond to several properties of the reducing theory. Here, it is the other way round: several properties of the reduced theory correspond to one property of the reducing theory, which means that the reduced theory does not supervene on the reducing theory. Again, the lack of the appropriate supervenience relation between T_{TD} and T_{BSM} is the culprit.

For the present purposes, this exposition of the threat DP poses for the foundational project must suffice. In the following, I will suggest an alternative method of carving up phase space that avoids DP while keeping the functionality and spirit of BSM intact. By avoiding DP, supervenience of T_{TD} on T_{BSM} will no longer be blocked and a fortiori, this threat to the foundational project can be avoided.

3.2 From Macro Regions to Worlds

Instead of disjoint macro regions as described above, I suggest employing an alternative way of carving up phase space—something that, to some extent, resembles macro regions, yet differs in important respects. In order to avoid the risk of confusion with ordinary macro regions, these new sets of phase space points shall be called *worlds*. They will be defined by a relation of *sufficient similarity* (SS), in a way to be made precise shortly. As a result, there no longer will be a partition into a finite number of macro regions of non-zero volume and with sharp boundaries, where different macro regions stand in a one-to-one correspondence to different macro states. Once the idea that we don't need such a partition has grown upon us, an adequate correspondence between the continuous evolution of the micro state and the thermodynamically expected, continuous evolution of the macro state can be recovered.

Contrary to the micro states contained in a standard macro region, the micro states in a world do not all instantiate the *same*, i.e., *identical* macro states. (Some do, namely those representing permutations of identical constituents.)[20] Instead, the construal of worlds is based on the notion of *indistinguishable* macro states, which defines a broader class: identical macro states are certainly indistinguishable, but indistinguishable macro states are not necessarily identical. This, to some, might seem as an unfavourable introduction of subjectivity[21] into the framework. Speaking of indistinguishable macro states triggers the question: indistinguishable for whom or what? However, replacing sameness with indistinguishability allows the resulting models to represent continuous macroscopic change, in accordance with the continuous evolution of the micro state, and with target systems for which a thermodynamic description is adequate. So there clearly are some advantages. In the end, it will be for the reader to decide whether they prefer a framework that contains no reference to (in)distinguishability, but is unable to model continuous macroscopic change, or a framework that is able to model continuous macroscopic change, but refers to (in)distinguishability.

Just how exactly this new framework is built and supposed to work is the topic of this section.

---

[20] For a discussion of the permutation of identical constituents/particles, see below.

[21] The term "subjectivity" might actually be a bit misleading. What I am going to provide is an objective description that takes into account matters of perspective. There is nothing subjective about this description itself. Cf. Hemmo and Shenker (2012), who emphasize that macro states, and hence partitioning into macro regions, are objective, despite being observer-relative, and Callender (1999, p. 370), pointing out that "the coarse graining used to define [Boltzmann entropy] does not imply that [Boltzmann entropy] is subjective, only that it is relational".

3.2.1 Introducing a New Framework for BSM: Some Conditions

Most importantly, by developing this new framework, I want to avoid the discontinuity problem as it arises from the old framework. To achieve this goal, my construal of worlds as sets of micro states in phase space should avoid the imposition of sharp boundaries between different worlds, each corresponding to a particular macro state, such that macro values jump discontinuously upon the micro state crossing such boundaries. That, however, is not to say that worlds may not have *any* boundaries at all. In fact, they may even have sharp boundaries. But these boundaries must differ from those we encounter in the old framework in that crossing them does not result in discontinuous macroscopic change. This will be made possible by allowing worlds to overlap, without micro states in the overlap instantiating inconsistent macro states, as is the case for overlapping macro regions in the old framework (see section *Removing no-overlap*). In order to avoid such inconsistencies, a micro state in an overlap must not instantiate *different* macro states. As shown above, micro states in overlapping macro regions would instantiate different macro states (which is why there can't be overlapping macro regions in the old framework). This is due to the fact that the finite set of different macro regions is bijectively mapped onto the finite set of different macro states: each macro region corresponds to exactly one macro state, and vice versa. So, to avoid inconsistent instantiations in the overlap of worlds, worlds must not correspond to exactly one macro state, and macro states must not correspond to exactly one world. Rather, worlds shall correspond to a set of macro states that are indistinguishable[22] from a given macro state. This set can be a continuum of macro states, which, as we will see, enables modelling continuous macroscopic change. Of course, the reasonable stipulation that every micro state instantiates exactly one macro state will not be touched, for this would undermine supervenience of the macro state on the micro state. Accordingly, since worlds are sets of micro states, and are supposed to correspond not to one, but a continuum of macro states, the micro states within a world cannot all instantiate the same macro state, as the micro states in macro regions do. Instead, micro states within a world will be related in that they all instantiate macro states that are indistinguishable from a given macro state. This feature will be expressed in the relation "sufficient similarity" (SS), pertaining to micro states: a world, then, will be defined as a set of sufficiently similar micro states, such that all micro states in a world instantiate a macro state that is indistinguishable from a given macro state.

It should be mentioned that already in the old framework, macro states trivially can be indistinguishable—namely if they are identical. Because of that, and because of how the relation SS is defined via indistinguishability, micro states in the old framework can be sufficiently similar as well. In fact, since all micro states in a macro region instantiate the same macro state, all micro states in a macro region are

[22] Relative to a given "resolution power", see below.

sufficiently similar to all other micro states in the same macro region. Here, we already encounter an important difference between worlds and macro regions. The micro states in a world are not necessarily sufficiently similar to all other micro states in that world. It is merely required that they be sufficiently similar to what I call the "central" micro state of a world, around which the world is defined. Take, for example, some arbitrary micro state. It certainly instantiates a macro state. If we want to construe a world around this micro state, we merely require that all micro states in this world be sufficiently similar to that micro state in that they instantiate a macro state that is indistinguishable from the macro state instantiated by the selected micro state. This does not entail the requirement that all micro states in the thusly construed world instantiate macro states indistinguishable from the macro states instantiated by all other micro states in that world. It is very much allowable that some micro states in a world instantiate macro states that aren't mutually indistinguishable. Hence, not all micro states in a world are necessarily sufficiently similar to all other micro states in the same world. At the same time, micro states from different worlds may be sufficiently similar in that they instantiate indistinguishable macro states. In other words: both relations, indistinguishability as well as sufficient similarity, are nontransitive.

Let's make things reasonably precise.

First, supervenience of the macro state on the micro state is claimed. That is, if the macro state changes, the micro state must change as well. In other words: every micro state instantiates exactly one macro state.[23] Formally:[24]

$$\forall x_i \exists! M_i : (M_i x_i) \qquad \text{(supervenience)}$$

Read: For every micro state x_i there is *exactly one* macro state M_i such that x_i instantiates M_i.

Like before, macro states are defined via sets of macro values. Saying that macro states are indistinguishable means that it is impossible to tell them apart based on these macro values. So the question arises: when, exactly, is it impossible to tell macro states apart based on their macro values (and when is it possible)? When are

[23] According to Frigg and Werndl (2011, p. 125), supervenience of the macro state on the micro state is "a basic posit of the Boltzmann approach". In this respect, the new framework doesn't differ from the old one. See also Frigg (2008, 3.2.1) and Werndl and Frigg (2015a, passim), and recall Frigg et al. (2016, 4.1) from above: "to every given microstate x there corresponds *exactly one* macrostate".

[24] $\exists!$ denoting unique existential quantification—"there is exactly one"—as opposed to normal existential quantification with the quantifier $\exists$—"there is (at least) one"—allowing for the possibility that there is more than one. Unique existential quantification is used for brevity. It can be expressed in terms of normal existential quantification: instead of formulating the supervenience condition as $\forall x_i \exists! M_i : (M_i x_i)$, one can equivalently write $\forall x_i \exists M_i : \left(M_i x_i \wedge \neg \exists M_j \left(M_j x_i \wedge M_i \neq M_j\right)\right)$.

macro values similar enough? In order to answer these questions, I must elaborate the way in which macroscopic distinguishability depends on "resolution power".[25]

Saying that macroscopic (in)distinguishability depends on resolution power means the following: observers and measuring instruments have certain limitations with respect to their capabilities to differentiate between different macro states. Depending on their abilities/precision, to which I refer as their resolution power, two given macro states, specified by different macro values, might be distinguishable for a given observer/instrument, or not. Thus, macroscopic (in)distinguishability is observer- or instrument-relative in that it depends on the abilities or limitations of the observer or instrument: relative$_{\text{observer}}$[26] to a given observer and/or instrument with a certain resolution power, certain macro states are distinguishable (or not). Accordingly, for indistinguishable macro states, it is not claimed that they are specified by identical macro values, as it is done for same macro states in the old framework. Indistinguishable macro states may differ in their macro values. For example, the macro states M_1 and M_2 might differ in temperature values $T_1 = 23.542°C$ and $T_2 = 23.543°C$. But with a limited resolution power on the order of $\Delta T = 0.1°C$, they are, with respect to temperature, rendered indistinguishable, despite being different. Already on a conceptual level, it seems reasonable to not regard *different* macro states as the *same* macro state, even if they are indistinguishable. Macroscopic indistinguishability being relative$_{\text{observer}}$ comes on top: in principle, a being like Laplace's Demon can distinguish between all possible micro states, and thus all macro states. And in the example given, even a human observer can render the macro states in question distinguishable, if they increase their resolution power, for example by using a better thermometer. Thus, it seems unreasonable to identify macro states based on their indistinguishability, which is a contingent, observer-relative, matter of fact. Nevertheless, macro states can be grouped together via their indistinguishability. For example, if one measures or prepares the temperature of a system to be $23.5°C$ in accordance with a certain resolution power, one may pretend that this is the precise macro value, namely 23.5000...°C. Thus, a precise macro state is specified, even though it is practically impossible to measure or control to absolute precision. Generally speaking, the *formal* specification of macro states via macro values is not limited by resolution power: one is free to be arbitrarily precise when specifying macro states. Irrespective of that, due to limited resolution power, other macro states, differing in macro values only to an extend that cannot be resolved, are rendered indistinguishable from the precise macro state specified. Via this indistinguishability, macro states can be grouped together with a given macro state, without identifying them with the latter, or with one another.

Having established that macroscopic (in)distinguishability is observer-relative, the question above can be answered: macro states are distinguishable iff, given a certain resolution power, an observer/instrument can distinguish between their respective macro values. For example, M_1 and M_2 might actually have different

[25] Borrowing the term from Hemmo and Shenker (2012), without necessarily meaning the exact same concept as they do.

[26] I write "observer" as a shorthand for "observer and/or instrument".

temperature macro values $T_1 = 23.542°\text{C}$ and $T_2 = 23.543°\text{C}$, but given a resolution power of $\Delta T = 0.1°\text{C}$, these macro values can't be distinguished through observation or measurement. In this sense, they are indistinguishable.

With the notion of macroscopic (in)distinguishability made clear, worlds can now be introduced into the framework. First, sufficient similarity between micro states is defined via macroscopic indistinguishability of macro states (where "$\Leftrightarrow$" denotes macroscopic indistinguishability):

$$\forall x_{i,j} : \left(x_j SS x_i \leftrightarrow \exists M_{i,j} \left(M_j x_j \wedge M_i x_i \wedge \left(M_j \Leftrightarrow M_i \right) \right) \right) \qquad \text{(sufficient similarity)}$$

Read: For all pairs of micro states $x_{i,j}$: x_j and x_i are sufficiently similar iff there are macro states M_i and M_j such that M_j is instantiated by x_j, M_i is instantiated by x_i, and M_j and M_i are indistinguishable.

It is then possible to construe a world $\mathcal{W}_i$ as the set of all x_j that satisfy SS with respect to a given x_i:[27]

$$\forall x_i \exists \mathcal{W}_i \left(\forall x_j \left(x_j \in \mathcal{W}_i \leftrightarrow x_j SS x_i \right) \right) \qquad \text{(world membership)}$$

Spelled out in terms of indistinguishable macro states, *world membership* reads:

$$\forall x_i \exists \mathcal{W}_i \left(\forall x_j \left(x_j \in \mathcal{W}_i \leftrightarrow \exists M_{i,j} \left(M_j x_j \wedge M_i x_i \wedge \left(M_j \Leftrightarrow M_i \right) \right) \right) \right)$$

For all micro states x_i there is a world $\mathcal{W}_i$ such that, for all micro states x_j, x_j is a member of $\mathcal{W}_i$ iff there are macro states M_i and M_j such that M_j is instantiated by x_j, M_i is instantiated by x_i, and M_j and M_i are indistinguishable.

Under this construal, micro states instantiating indistinguishable macro states belong to the same world.

A few remarks are in order before moving on. First, it is worth stressing that SS is a reflexive and symmetric but nontransitive relation. It inherits these relational properties from the notion of macroscopic indistinguishability, itself being nontransitive. Hence, worlds, as defined via SS, are not equivalence classes, as disjoint

---

[27] One might think that defining worlds *around* a central micro state is somewhat odd. But the idea is not at all new. Take Reichenbach (1956, p. 71) for an example, even calling this method "natural": "The distinction between microstates and macrostates [...] leads to a natural grouping of microstates. Given a microstate a_1, the macrostate A to which it belongs is well determined: A includes all those microstates a_i which are macroscopically indistinguishable from a_1." Note the slight differences in terminology. What Reichenbach calls a "macrostate" is a set of micro states and thus would be called a *macro region* in the old framework. But actually, since it is only required that the a_i be macroscopically indistinguishable from a given a_1, i.e., that they satisfy SS with respect to a_1, the resulting set is not a *macro region*, but a *world* in our sense: not all a_i necessarily are macroscopically indistinguishable on Reichenbach's construal. Unfortunately, Reichenbach didn't seem to realize that, once worlds—macrostates in his terminology—are defined around a given micro state via a similarity relation, it is only a small step towards overlapping worlds.

macro regions in a partition are. While all micro states x_j in a world $\mathcal{W}_i$ are sufficiently similar to the central micro state x_i, it is not necessary that all x_j in said world *mutually* satisfy SS.[28]

Second, x_i is called the central micro state of the world $\mathcal{W}_i$. In this vein, every micro state is the central micro state of its own world, and conversely, every world has a central micro state. One might want to know how to find the central micro state and its world, given only a macro state, specified via certain macro values. Practically, due to macro states being distinguishable relative$_{\text{observer}}$ to a certain resolution power, this is impossible, but also unnecessary. One can easily specify a macro state to arbitrary precision, simply by fixing macro values. By contrast, it is not at all easy to specify a precise micro state: one would have to fix the values of all (relevant) degrees of freedom of all constituents of the system of interest in order to specify a unique point in phase space. This is unfeasible. The best one can do practically is to prepare a system such that it is in a macro state indistinguishable from the specified macro state, according to a given resolution power. The actual micro state of the system so prepared is then sufficiently similar to other micro states which, counterfactually, instantiate macro states that are indistinguishable from the specified one. Alternatively, one can measure the macro values of the system and take them as the specification of the macro state. The system will certainly be in some micro state, instantiating a macro state indistinguishable from the one specified. For all practical purposes, this is enough. If one wants to close in on a specific micro state, the preparation of the system must become better, or resolution power must be increased, such that the world of sufficiently similar micro states becomes smaller. But since practically, resolution power is limited, it is practically unattainable to find a specific micro state this way.

Let's move on with building the framework. Earlier, for the old framework, we have seen that regions of phase space not corresponding to any macro state—what I have called non-regions—are rather absurd: every point in phase space precisely describes a possible micro state and thus, via *supervenience*, (counterfactually) instantiates a macro state. But as soon as a micro state instantiates a macro state, it is part of a macro region (in the old framework), or of a world (in the new framework). Hence, every possible micro state is part of a world. This is called *exhaustive cover*:

---

[28] For an illustrative example, think of the height of people: When Amy and Berta are of the same height, and Berta and Chris are of the same height, then necessarily Amy and Chris are of the same height. Sameness is a reflexive, symmetric and transitive relation. Not so similarity. If Amy and Berta are of similar height, and Berta and Chris are of similar height, then it could be that Amy and Chris are of similar height, but it is by no means necessary. Hence, other than sameness, similarity is reflexive and symmetric but nontransitive. Likewise, the case of macroscopic indistinguishability: even if I am unable to notice the difference in temperature of a gas at different times t and t′, and unable to notice the difference in temperature of the same gas at t′ and t″, I might be able to notice the difference in temperature of the gas at t and the gas at t″. Hence, despite the gas at t and t′ being macroscopically indistinguishable (with respect to temperature), and the gas at t′ and t″ being macroscopically indistinguishable, it does not follow that the gas at t and t″ is macroscopically indistinguishable.

$$\forall x_i \exists \mathcal{W}_i (x_i \in \mathcal{W}_i) \qquad \text{(exhaustive cover)}$$

Exhaustive cover immediately follows from *world membership*, *sufficient similarity* and *supervenience*: take *world membership* and set $x_j = x_i$, such that $\forall x_i \exists \mathcal{W}_i (x_i \in \mathcal{W}_i \leftrightarrow x_i SS x_i)$. That $x_i SS x_i$ is satisfied can be seen from its analysis in *sufficient similarity*: $\forall x_i : (x_i SS x_i \leftrightarrow \exists M_i (M_i x_i \wedge (M_i \Bumpeq M_i)))$. That, for every x_i, there is an M_i such that it is instantiated by x_i is seen from *supervenience*, and $M_i \Bumpeq M_i$ is trivial. So $\forall x_i \exists M_i (M_i x_i \wedge (M_i \Bumpeq M_i))$ is satisfied. So x_i is sufficiently similar to itself: $x_i SS x_i$. And since the quantification ranges over all x_i, all x_i are members of the World $\mathcal{W}_i$ around x_i, i.e., of their own world. Thus, phase space is exhaustively covered by worlds.

However, this is not to say that every micro state belongs to only one world. Via *world membership*, spelled out in terms of macroscopic indistinguishability, all micro states instantiating macro states indistinguishable from a given macro state belong to the same world. Phase space is covered by worlds, and every micro state is a member of its own world. With these worlds also containing other micro states, it immediately follows that worlds overlap. The possibility of a micro state being a member of different worlds is one of the discerning features of the new method of carving up phase space. Iff a micro state x_k belongs to worlds $\mathcal{W}_i, \mathcal{W}_j$, it lies in their overlap $\mathcal{O}_{\mathcal{W}_{i,j}} = \mathcal{W}_i \cap \mathcal{W}_j$:

$$\forall x_k \forall \mathcal{W}_{i,j} \left(x_k \in \mathcal{O}_{\mathcal{W}_{i,j}} \leftrightarrow x_k \in \mathcal{W}_i \wedge x_k \in \mathcal{W}_j \right) \qquad \text{(overlap)}$$

Like *world membership* before, *overlap* is spelled out in terms of indistinguishable macro states:

$$\forall x_{i,j,k} \forall \mathcal{W}_{i,j} : \left(x_k \in \mathcal{O}_{\mathcal{W}_{i,j}} \leftrightarrow \exists M_{i,j,k} \left(M_k x_k \wedge M_i x_i \wedge M_j x_j \wedge (M_i \Bumpeq M_k) \wedge (M_j \Bumpeq M_k) \right) \right)$$

From this, it can be easily seen that, if there are micro states x_i, x_j and x_k, instantiating—via *supervenience*—macro states M_i, M_j and M_k, and M_k is indistinguishable from both M_i and M_j, then there is an overlap of worlds $\mathcal{W}_i$ and $\mathcal{W}_j$ containing the micro state x_k. It is important to note that for this result, it is not necessary to assume that M_i and M_j are indistinguishable.

The converse can be seen just as easily: if there is an overlap containing an x_k, then there is a macro state M_k that is indistinguishable from both M_i and M_j, even if the latter two are distinguishable.

Figure 3.16 illustrates these results. Two micro states, x_i and x_j, instantiate distinguishable macro states, M_i and M_j. One can think of this picture as superimposed snapshots of the micro evolution of a system, at different times. Say, for example, that x_i is the micro state of the system at time t, and that x_j is the micro state of the system at a later time t''. The picture represents a case in which our resolution power is good enough to distinguish between the macro states instantiated by x_i at t and x_j at t''. Accordingly, x_i is not in $\mathcal{W}_j$, which would render it sufficiently similar to x_j, and likewise, x_j is not in $\mathcal{W}_i$. During its evolution from x_i to x_j in the time interval $[t,t'']$, the micro state traverses the overlap $\mathcal{O}_{\mathcal{W}_{i,j}}$. Let's pick out x_k as the micro state at time t', between t and t''. Any x_k in the overlap will do. Since it is in the overlap, it instantiates a macro state M_k, that, with the given resolution power, is indistinguishable from both, M_i and M_j.

The phase points x_i, x_j and x_k in Fig. 3.16 are drawn at a significant distance, because they each represent the micro state of a system at different times t,t',t''. The picture drawn is an overlay of "snapshots" at these times. But it is easy to imagine the continuous trajectory of the actual evolution of the micro state through phase space during the time interval $[t,t'']$ instead. This trajectory being continuous—as is assumed for the micro-state evolution in the old as well as in the new framework—contains an uncountable infinity of micro states. And since all these micro states instantiate a macro state, the uncountably infinite set of micro states in the trajectory bijectively corresponds to an uncountably infinite set of macro states instantiated, and thus, to an uncountably infinite set (of sets) of macro values bijectively corresponding to the set of macro states. So the issue that gave rise to the discontinuity problem in the old framework—the issue that there was an uncountable infinity of micro states that couldn't be mapped bijectively onto a finite set of

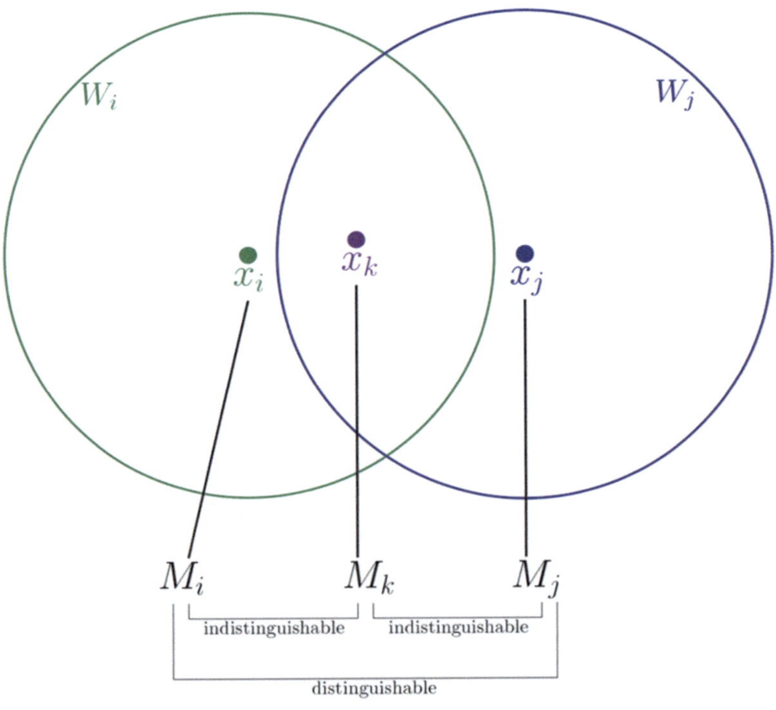

Fig. 3.16 Three micro states at different times. While the first and second macro states instantiated are mutually indistinguishable, and so are the second and the third, the first and third macro states are distinguishable

macro states—seems to have vanished completely. In the new framework just proposed, we likewise want to be able to speak of entropy increase, so we likewise have to have worlds of different volumes, so we likewise have to have worlds of non-zero volume. But by contrast, models in the new framework aren't restricted to a finite number of worlds by these requirements, because there are as many worlds as there are micro states, and there are as many worlds as there are micro states—despite worlds having non-zero volume—because worlds can overlap, and worlds can overlap because the micro states they contain don't all instantiate the *same* macro state, but merely ones that are indistinguishable from a given macro state. Because of all this, change of macro values doesn't have to be modelled as discontinuous in the new framework, as opposed to the old one. By using macroscopic indistinguishability instead of sameness, it is possible to avoid models in which the evolving micro state, in order to instantiate different macro states, must cross a boundary at which macro values jump. In models resulting from the new framework, the micro state can wander through phase space without crossing such boundaries, instantiating indistinguishable macro state after indistinguishable macro state, until, at the two far ends of its continuous trajectory, it does instantiate distinguishable macro states.

One might be tempted to think that accounting for limited resolution power would solve the discontinuity problem already in the old framework: simply make the partition so fine-grained that neighbouring macro regions correspond to different, yet indistinguishable macro states. Then, the transition of the micro state from one to the other wouldn't result in a noticeable discontinuity. This latter claim, speaking of *noticeable* discontinuities, is correct. However, it doesn't solve the discontinuity problem. DP doesn't arise merely because of *noticeable* discontinuities, but because of discontinuities *simpliciter*, whether noticeable or not; because the old framework cannot possibly model continuous change of macro values. It can't because there is only ever a finite set of allowable macro values in a given model, and a finite set of macro values cannot possibly be mapped bijectively onto a continuum of macro values as present in thermodynamically described target systems, and likewise, onto a continuum of micro states in a trajectory through phase space. This remains an idealisation—no matter how fine-grained the partition—that is problematic especially when regarding BSM as a fundamental theory (see above). By contrast, the new framework is able to model continuous change of macro values because it actually allows for a continuum of macro values to exist in its models, and this continuum can be mapped bijectively onto a continuum of micro states in a trajectory, and, trivially, onto a continuum of macro values as present in target systems that are adequately described by thermodynamics.

In short, as long as the micro states in a macro region are all supposed to instantiate the same macro state, the discontinuity problem arises. The new framework doesn't solve it merely by taking into account resolution power, but by replacing macro regions with worlds, sets of micro states that instantiate macro states *indistinguishable* from a given macro state, but not necessarily the same, such that worlds can correspond to a continuum of macro values. Indistinguishability between macro states alone doesn't imply that a set of macro values is a continuum. But at least, the new framework is able to model such cases. In solving the discontinuity problem, all the crucial work is done by replacing a finite set of macro regions with an uncountable infinity of worlds.

So much for a general overview of the proposed framework. In the following section, I will demonstrate its capabilities by applying it to the example of a gas spreading out in a room, familiar from above.

3.2.2 *Back to the Example*

Suppose there is an initial micro state x_0 instantiating M_0 with the gas concentrated in the corner of the room. The set of sufficiently similar micro states, via *world membership*, forms the world $\mathcal{W}_0$ around x_0. Let x evolve away from x_0, such that the gas is spreading out: x_1 instantiates a slightly, but noticeably more spread out macro state of the gas, M_1. Since M_1 is macroscopically distinguishable from M_0, x_1 is not a member of $\mathcal{W}_0$. But it certainly is a member of its own world, $\mathcal{W}_1$. Despite x_1 and x_0 residing in different worlds $\mathcal{W}_0$ and $\mathcal{W}_1$, we can select x_1 such that their worlds do overlap: there are (uncountably many) micro states between x_0 and x_1 that meet the *world membership* condition of both, $\mathcal{W}_0$ and $\mathcal{W}_1$. They make up the overlap $\mathcal{O}_{\mathcal{W}_{0,1}}$. Let's pick out one of them, $x_{\mathcal{O}}$. The macro state $M_{\mathcal{O}}$ it instantiates is indistinguishable from M_0, such that $x_{\mathcal{O}}$ is part of $\mathcal{W}_0$. But $M_{\mathcal{O}}$ is also indistinguishable from M_1, such that $x_{\mathcal{O}}$ is part of $\mathcal{W}_1$. So, when x evolves from x_0 to $x_{\mathcal{O}}$, and hence the instantiated macro state changes from M_0 to $M_{\mathcal{O}}$, this change is unnoticeable. When x then moves from $x_{\mathcal{O}}$ to x_1, the macro state changes from $M_{\mathcal{O}}$ to M_1, but again, this change is unnoticeable. Yet now, x instantiates M_1, which is noticeably different from M_0. So, through a series of unnoticeable changes, the macro state has changed noticeably. In this fashion, we can describe the gas in its evolution from residing in the corner to being spread out over the whole room, through a series of unnoticeable changes that together account for the very noticeable difference between the initial state and the equilibrium state. And remember that the micro states we are referring to in this example are merely selected snapshots from a continuous trajectory, containing an uncountable infinity of actual micro states, instantiating an uncountable infinity of macro states. From such an uncountably infinite set of micro states, we can always select a finite set of three micro states that behave in the way just described for x_0, $x_{\mathcal{O}}$ and x_1.

3.2.3 *World Size, the Equilibrium World, and Entropy Increase*

So much for the continuous change between macro states. But how about the sizes of worlds? In particular, is $\mathcal{W}_{eq}$, like $\Gamma_{M_{eq}}$, still dominant, i.e., "vastly larger than any other [world]", such that the accessible phase space volume is "almost entirely taken up by equilibrium microstates"? Given the special interest in M_{eq}, will x_{eq} and its world $\mathcal{W}_{eq}$ turn out to be special? Can the approach to and remaining in this special world still be explained in this framework? Is the change of entropy over time now continuous? A few considerations are important.

First, we must state clearly what determines Boltzmann entropy. In the old framework, it was assumed that all micro states in a macro region instantiate the same macro state. Endowing phase space with a suitable measure for macro regions (Frigg & Werndl, 2011, p. 125f.), "the Boltzmann entropy of a macro state M [...] measures the portion of the system's γ -space [i.e., phase space] that is taken up by microstates that correspond to M". In other words, it depends on the volume of the macro region corresponding to a given macro state. Thus, Boltzmann entropy can be assigned directly to a macro state (Frigg & Werndl, 2012, p. 102): "The Boltzmann entropy of a macrostate M_i is $S_B(M_i) := k_B \log\left[\mu\left(\Gamma_{M_i}\right)\right]$ (where k_B is the Boltzmann constant), and the Boltzmann entropy of a *system* at time t, $S_B(t)$, is the entropy of the macrostate of the system at t: $S_B(t) := S_B\left(M_{x(t)}\right)$, where $x(t)$ is the microstate at t and $M_{x(t)}$ is the macrostate corresponding to $x(t)$."

For the new framework, we can retain the latter definition of the entropy of a system at time t, but the definition of the entropy of a macro state must be slightly altered: it must depend on the volume of *worlds*, not on the volume of macro regions. Accordingly, Boltzmann entropy in the new framework is assigned to a given macro state M_i by considering not only M_i itself, but all macro states that are indistinguishable from M_i. This means that we have to take into account the set of all micro states that instantiate macro states indistinguishable from M_i, i.e., the world $\mathcal{W}_i$: $S_{B(new)}(M_i) := k_B \log\left[\mu(\mathcal{W}_i)\right]$.

It is a distinctive feature of the new framework with overlapping worlds that the sum of the volume of all worlds is infinite. Since every micro state has a world of non-zero volume around it, and there are infinitely many micro states in a given, finite volume of available phase space, there are infinitely many worlds of non-zero volume in a finite volume of phase space. But this does not impede the possibility of comparing the sizes of individual worlds with each other, or with the volume of the available phase space. For example, for a particular equilibrium macro state, we should still be able to say that the world consisting of all micro states instantiating macro states indistinguishable from it is larger than worlds consisting of micro states instantiating macro states indistinguishable from a given non-equilibrium macro state, and thus assign a higher entropy value to the former. Likewise, since the volume of worlds remains comparable to the volume of the available phase space in total, we should still be able to say that, for example, a particular equilibrium world covers almost all of the available phase space volume, while other, non-equilibrium worlds, don't.

Note that the equilibrium world $\mathcal{W}_{eq}$ is defined around the micro state x_{eq} that instantiates the equilibrium macro state M_{eq}. There is nothing that enforces the selection of this particular equilibrium micro state, or this particular equilibrium macro state. After all, there is a continuum of them. As long as a micro state instantiates a macro state indistinguishable from equilibrium, any will do. Nevertheless, we might be able to search for the "best" x, if we want or need to. Just select one micro state that instantiates a macro state indistinguishable from equilibrium, and then shift it around, until, via *world membership*, the volume of its world is maximized. This way, a unique "best" x and a unique equilibrium world is selected. The thusly selected equilibrium micro state may be called the "true" equilibrium micro state. Of course, whether the actual micro state of a system in macroscopic equilibrium is precisely the true equilibrium micro state (with almost certainty, it isn't) or just one sufficiently similar is irrelevant for all practical purposes.

Lastly, approach to and remaining in equilibrium for long time periods have essentially the same explanation in the new framework as in the old: because of their relative sizes, the micro state is more likely to evolve from smaller worlds into larger ones, and ultimately to end up in an equilibrium world, where it tends to stay. It goes without saying that, nevertheless, fluctuations out of equilibrium are still possible, despite being incredibly unlikely. Yet, the new framework differs in this respect, too. Remember the issue of discontinuous Boltzmann entropy increase addressed above: with the micro state moving from one macro region to another, both being differently sized, the entropy value changes discontinuously. Now, for the new framework, this is not necessarily so: whether Boltzmann entropy changes continuously or not, together with the continuous evolution of the micro state, depends on whether the volume of the world actualized changes continuously or not. The new framework allows for this possibility, by contrast to the old framework, where Boltzmann entropy change must be modelled as discontinuous.

3.3 Conclusion

I have presented the discontinuity problem of the standard framework of BSM, and argued that an alternative way of partitioning phase space into overlapping worlds is called for, given one wants to avoid it: on the hitherto usual method of partitioning phase space into finitely many macro regions of non-zero volume, macro values are always modelled to change in a discontinuous fashion, if they change at all. This is at odds with both the continuous evolution of the micro state, and the expectation that a framework employed for modelling classical thermodynamic phenomena should be able to model continuous change of macro values.

I tried to fix this issue by excising *no-overlap* and *exhaustive-cover* from the old framework. This didn't bring about the results we had hoped for. Removing the latter resulted in "non-regions", which subsequently turned out to be ordinary macro regions after all. But even if one insists that they are not ordinary, the problem of discontinuous change upon the micro state crossing boundaries between macro regions remains. Removing no-overlap, on the other hand, resulted in the micro state instantiating inconsistent macro states at the same time.

Hence, I proposed an alternative framework, featuring overlapping worlds as a replacement for macro regions, and allowing for macro states to be indistinguishable despite being different. The key features of the new framework made it possible to have micro states, residing in the overlaps of worlds, instantiating macro states indistinguishable from macro states that themselves are mutually distinguishable. In this fashion, the macro state of a system can change continuously, without jumps between macro values, and in accordance with the continuous evolution of the micro state. The framework presented, it seems, is indeed a viable alternative to the old one, solving a central, albeit hitherto largely overlooked problem of the latter.

It goes without saying that much more work needs to be done if one really wants to establish the new framework as a superior alternative to the standard framework. So far, it is merely a proposal, sketched in broad strokes. To develop it into a full theory, its details need to be worked out. That, however, lies beyond the scope of this book. The primary objective I pursued with this chapter was to introduce and develop the concept of overlapping worlds. This will be needed in the following chapter, where I provide an account of how so-called world histories are formed from the discontinuous actualizations of universal particle configurations in Tychistic Bohmian Mechanics. There, what I have called the discontinuity problem arises even more severely: how can *discontinuous* actualizations of the configuration of the universe be reconciled with the seemingly *continuous* histories we experience? To do this, I propose to employ the notion of overlapping worlds. My reason to first introduce and develop this concept for BSM is simple: if I want to use a certain concept in order to reconcile an actually discontinuous micro dynamics—and thus an actually discontinuous macro dynamics—with seemingly continuous phenomena, this concept should also be employable to reconcile an actually continuous micro dynamics with what not only seems to be continuous phenomena, but, in classical thermodynamics, are assumed to be continuous phenomena. I have used BSM as a test bed for the concept of overlapping worlds, to be applied to the case of TBM's discontinuous actualizations in the following chapter.

References

Albert, D. Z. (2000). *Time and chance*. Harvard University Press.

Batterman, R. (2020). Intertheory relations in physics. In E. N. Zalta (Ed.). *The Stanford Encyclopedia of Philosophy* (Fall 2020 Edition). https://plato.stanford.edu/archives/fall2020/entries/physics-interrelate/

Callender, C. (1999). Reducing thermodynamics to statistical mechanics: The case of entropy. *The Journal of Philosophy, 96*(7), 348–373.

Dizadji-Bahmani, F., Frigg, R., & Hartmann, S. (2010). Who's afraid of Nagelian reduction? *Erkenntnis, 73*(3), 393–412. https://doi.org/10.1007/s10670-010-9239-x

Ehmann, A. (2022). Reducing thermodynamics to Boltzmannian statistical mechanics: The case of macro values. *Synthese, 200*, 513. https://doi.org/10.1007/s11229-022-03985-4

Fletcher, S. C. (2019). On the reduction of general relativity to Newtonian gravitation. *Studies in History and Philosophy of Modern Physics, 68*, 1–15. https://doi.org/10.1016/j.shpsb.2019.04.005

Frigg, R. (2008). A field guide to recent work on the foundations of statistical mechanics. In D. Rickles (Ed.), *The Ashgate companion to contemporary philosophy of physics*. Ashgate.

Frigg, R., Berkovitz, J., & Kronz, F. (2016). The Ergodic hierarchy. In E. N. Zalta (Ed.). *The Stanford Encyclopedia of Philosophy* (Summer 2016 Edition). https://plato.stanford.edu/archives/sum2016/entries/ergodic-hierarchy

Frigg, R. & Hartmann, S. (2020). Models in science. In: E. N. Zalta (Ed.). *The Stanford Encyclopedia of Philosophy* (Spring 2020 Edition). https://plato.stanford.edu/archives/spr2020/entries/models-science/

Frigg, R., & Werndl, C. (2011). Entropy: A guide for the perplexed. In C. Beisbart & S. Hartmann (Eds.), *Probabilities in physics* (pp. 115–142). Oxford University Press. https://doi.org/10.1093/acprof:oso/9780199577439.001.0001

Frigg, R., & Werndl, C. (2012). A new approach to the approach to equilibrium. In Y. Ben-Menahem & M. Hemmo (Eds.), *Probability in physics* (pp. 99–113). Springer Science & Business Media.

Frigg, R., & Werndl, C. (2019). Statistical mechanics: A tale of two theories. *The Monist, 102*(4), 424–438. https://doi.org/10.1093/monist/onz018

Giancoli, D. C. (2010). *Physik* (3rd ed.). Pearson.

Goldstein, S., Huse, D. A., Lebowitz, J. L., & Tumulka, R. (2017). Macroscopic and microscopic thermal equilibrium. *Annalen der Physik, 529*, 1600301. https://doi.org/10.1002/andp.201600301

Goldstein, S., Lebowitz, J. L., Tumulka, R., & Zanghì, N. (2019). Gibbs and Boltzmann Entropy in Classical and Quantum Mechanics. *arXiv.org*. https://arxiv.org/abs/1903.11870v2

Hemmo, M., & Shenker, O. R. (2012). *The road to Maxwell's Demon*. Cambridge University Press.

Kemeny, J. G., & Oppenheim, P. (1956). On reduction. *Philosophical Studies, 7*(1), 6–19. https://doi.org/10.1007/BF02333288

McLaughlin, B. & Bennett, K. (2018). Supervenience. In: Zalta, E. N. (ed.). *The Stanford Encyclopedia of Philosophy* (Winter 2018 Edition). https://plato.stanford.edu/archives/win2018/entries/supervenience/

Norton, J. D. (2012). Approximation and idealization: Why the difference matters. *Philosophy of Science, 79*(2), 207–232. https://doi.org/10.1086/664746

Palacios, P. (2019). Phase transitions: A challenge for intertheoretic reduction? *Philosophy of Science, 86*(4), 612–640. https://doi.org/10.1086/704974

Reichenbach, H. (1956). *The direction of time*. Dover.

Robertson, K. (2020). Asymmetry, abstraction, and autonomy: Justifying coarse-graining in statistical mechanics. *The British Journal for the Philosophy of Science, 71*(2), 547–579. https://doi.org/10.1093/bjps/axy020

Sklar, L. (1993). *Physics and chance: Philosophical issues in the foundations of statistical mechanics*. Cambridge University Press.

Stone, J. (2018). *Belgium and the Netherlands swap land to change their national borders*. https://www.independent.co.uk/news/world/europe/belgium-netherlands-national-border-change-meuse-vise-eijsden-maastricht-a8141166.html Last accessed December 6, 2020.

van Riel, R., & Van Gulick, R. (2019). Scientific Reduction. In: E. N. Zalta, (Ed.). *The Stanford Encyclopedia of Philosophy* (Spring 2019 Edition). https://plato.stanford.edu/archives/spr2019/entries/scientific-reduction/

Wallace, D. (2018). *The necessity of Gibbsian statistical mechanics*. In submission. https://dornsife.usc.edu/assets/sites/1045/docs/gibbsboltzmann.pdf

Werndl, C., & Frigg, R. (2015a). Reconceptualising equilibrium in Boltzmannian statistical mechanics and characterising its existence. *Studies in History and Philosophy of Modern Physics, 49*, 19–31. https://doi.org/10.1016/j.shpsb.2014.12.002

Werndl, C., & Frigg, R. (2015b). Rethinking Boltzmannian Equilibrium. *Philosophy of Science, 82*(5), 1224–1235. https://doi.org/10.1086/683649

Wimsatt, W. C. (2006). Reductionism and its heuristics: Making methodological reductionism honest. *Synthese, 151*(3), 445–475. https://doi.org/10.1007/s11229-006-9017-0

Chapter 4
Worlds in a Stochastic Universe

All that has been said so far can be considered preliminaries for what follows. In the previous chapter, I have argued that, from the way the usual framework of BSM is set up, the discontinuity problem arises, and that it can be tackled by adopting an alternative framework, allowing for phase space to be carved up into a continuous set of overlapping worlds rather than a finite, discontinuous set of disjoint macro regions. The framework thus construed allows for modelling continuous change of the macro state of a system, in accordance with the continuous evolution of its micro state, and the expectation (from thermodynamics) that the evolution of macro values at least possibly should be continuous, all while leaving the functionality and spirit of BSM intact.

A similar, despite different problem arises for TBM. Here it is not the continuous evolution of the micro state that is at odds with the unexpected, discontinuous evolution of the macro state. Rather, the evolution of the micro state in TBM, i.e., the evolution of the universal particle configuration Q, is always discontinuous: the configuration randomly jumps through configuration space, thereby instantiating different worlds, one at a time. Successive actualizations of Q at different times are not linked up by (continuous) trajectories. Thus, successive worlds aren't linked up either. On this rests Bell's charge against his Everett (?) theory, and likewise against TBM: if worlds—i.e., configurations in his terminology—at different times aren't linked up, and only actualized for an instant, then reference to the past as well as to the future, from within such a world, is empty. Bell (1987a, 135–136): "there is no association of the particular present with any particular past. [...] Everett's replacement of the past by memories is a radical solipsism—extending to the temporal dimension the replacement of everything outside my head by my impressions".

Bell has a point. It is difficult to explain the common appearance of history evolving consistently over time based on individual, mutually independent, internally static configurations. Barbour, as mentioned earlier, has attempted to provide such an explanation, employing his concept of time capsules. However, as I will

A. Ehmann, *Worlds in a Stochastic Universe: On the Emergence of World Histories in Tychistic Bohmian Mechanics*, Synthese Library 537,
https://doi.org/10.1007/978-3-032-14632-8_4

argue, it is not easy to reconcile this view with our current best understanding of how appearances arise. Simply put, our current best theories suggest that any experience supervenes on processes happening in the brain, rather than from static, unchanging states. This is at odds with Barbour's proposal, for his time capsules are such static states, and he claims that the experience of history unfolding is due to the brain comparing snapshots from within such static states. To reconcile FAPP-continuous macroscopic change—the history of the universe as we perceive it, unfolding consistently—with randomly actualized configurations at the fundamental level of description, is the problem I am going to tackle in this chapter. That is, we are trying to reconcile our manifest image (Sellars) of how history unfolds with the scientific image presented by TBM. For this, a suitable link between configurations must be provided, despite their stochastic dynamics. Exploiting the notion of worlds as developed for solving the discontinuity problem of BSM above will prove extraordinarily helpful in this endeavour.

In the chapter introducing TBM, I have presented the solution Dürr and I have chosen in our recent paper, based on Barbourian time capsules and FAPP-classical world branches. Neither in this book so far, nor in our paper has this been spelled out in great detail. In the following, I am going to present my critique of the time-capsule approach, arguing that it is not necessary to account for macroscopic appearances, in particular macroscopic change, from within individual, static configurations, and that it faces further difficulties. At the same time, I will give a detailed account of how FAPP-continuous, historically consistent world histories can emerge from TBM's stochastic and discontinuous micro dynamics, in accordance with the branching structure of the WF.

4.1 Worlds from the Wave Function vs. Worlds from the Particle Configuration

Following the usual presentation in the Many-Worlds community,[1] worlds can be identified with decoherence-induced, quasi-classical branches of the WF. Dürr and I have adopted this notion for our approach. However, to some this approach might look a bit unnatural, given that TBM, as a Bohmian theory, is (at least something similar to) a primitive ontology theory,[2] with the Bohmian particles as the fundamental building blocks of material objects. Hence, it could be argued, the WF should not play the central role in defining the notion of a world within the TBM frame-

[1] Of which Wallace, in particular with his (2012), is one of the most prominent representatives.

[2] With primitive ontology theories being theories that provide a realist reading of some (value definite, usually with respect to position) variable—the so-called primitive variable—of their formalism that refers to the fundamental stuff in physical space or space-time that makes up the world and everything in it. See Allori (2015) for a review of the approach, and my discussion above.

work. Instead, the notion should be defined in terms of the primitive of the theory—the "empirical substructure" in van Fraassen's (1980, passim) terminology.

However, it is to be feared that an argument brought forward against standard BM[3] carries over to TBM: it can be argued that the particle configuration in standard BM is superfluous theoretical structure. If this holds for TBM, particles and their configurations should not be the basis upon which the notion of worlds is defined, for superfluous structure can be easily excised, thereby rendering said definition of worlds meaningless, and a fortiori undermining the concept that is supposed to account for the appearance of a consistent history in TBM. Let's assess this argument, headed 'Many worlds in denial?', by revisiting Valentini's (2010) and Brown's (2010) exchange.

4.1.1 Many Worlds in Denial?

This argument from the particle configuration being superfluous theoretical structure has to do with the ontological roles ascribed to the two parts of the complete state description in BM. The latter is often read realistically not only with respect to the configuration, but also with respect to the WF. Recall the discussion of the two-space reading above: because the WF, together with the particle configuration, is part of the complete state description and indispensable for the dynamics as well as the initial distribution of the latter, it should represent something physical. On this basis, the WF is interpreted as "an ontological—that is, physically real—complex-valued field $\Psi(q,t)$ on configuration space" (Valentini, 2010, 479).[4] And because of this being "ontological" of the WF, one could be lead to believe that BM is Many Worlds in denial (ibid.): "If one takes pilot-wave theory seriously as a possible theory of the world [i.e., the universe in our terminology], and if one thinks about it properly and carefully, one ought to see that it really contains many worlds—with a superfluous configuration q appended to one of those worlds."[5] Even unoccupied branches/wavepackets, according to this argument, should refer to something physical. Take, for example, two wave packets Ψ_1 and Ψ_2, resulting from a branching process of an original Ψ. They are said to not differ in terms of physicality (ibid., 493): "the motion of Ψ_1 represents a world every bit as bona fide as the world represented by Ψ_2". Valentini (ibid., 495, 497) argues to the contrary: since there really

[3] Cf. Valentini (2010) and Brown (2010).

[4] Note that Valentini bases his discussion on de Broglie's original pilot-wave theory, which doesn't speak of a quantum potential, and in this respect is closer in spirit to what is nowadays called Bohmian Mechanics than Bohm's original theory (see the distinction above). The latter was formulated by Bohm much later. Most importantly, however, Valentini is committed to an ontologically thick reading of the WF, as opposed to other proponents of BM.

[5] Note that this and the following quotation isn't Valentini arguing against BM, but his presentation of the argument brought forward against BM.

is only one "true history", i.e., only one wave packet endowed with the particle configuration, and it can be decided within pilot-wave theory which one that is, it is clear that there is only one real world and "the alternative trajectories are mathematical, not ontological." Hence, following Valentini, BM is not many worlds in denial.

As Brown (2010, 510 ff.) notes in his reply to Valentini, it is not necessary to subscribe to such a "dualist" particle-wf-ontology in the first place. In fact, one might be better off by adopting the Everett interpretation and not buying into the Bohmian particle ontology at all. On that, Brown is right: at least prima facie, one doesn't have to adopt the Bohmian ontology. However, for our purposes, i.e., for answering the question whether the configuration is superfluous in TBM, it is inadmissible to take over Brown's verdict regarding standard BM and straightforwardly apply it to the case of TBM. Also, even if Brown is right and BM indeed is many worlds in denial, such that the particle configuration is rendered superfluous, it is questionable whether this carries over to TBM. Both, BM and TBM, can feature a realist interpretation of the WF. And insofar a realist interpretation of the WF invariably renders the configuration superfluous, TBM falls prey to this objection as well. However, there are different ways in which things can be said to be real. In other words: there are variants of realist interpretations of the WF, and presumably, it depends on the variant whether the argument applies.

So, what precisely does a "realist interpretation" of the WF amount to in BM and TBM, respectively? Valentini likens the WF in BM to the electromagnetic field in Maxwell's formulation of electromagnetism (Valentini, 2010, 480): "This provides a particularly good parallel with pilot-wave theory. A charged test particle, placed in an external electromagnetic field $\mathrm{E}(\mathrm{x},t),\mathrm{B}(\mathrm{x},t)$, will follow a trajectory $\mathrm{x}(t)$. One would normally say that the field is real, and that the realized particle trajectory is real; while the alternative particle trajectories (associated with alternative initial positions $\mathrm{x}(0)$) are not real." The viability of such a realist reading of the electromagnetic field in Maxwell's formulation of electromagnetism notwithstanding, this suggests that the WF is viewed as being ontologically on a par with the particle configuration: both are regarded as equally real, just like particles and the electromagnetic field are regarded as equally real.

Brown also regards the WF as real, but in a particularly strong sense. He seems to view the WF as a kind of substance, or at least as something sufficiently substance-like, in the sense of a bearer of real, physical properties. Brown & Wallace (2005, 525) emphasize this point in response to Holland: "Quantum mechanics, in our view, has both substance and form before the introduction of hidden corpuscles."

As Brown (2010, 513) writes, in order to avoid the many-worlds-in-denial attack, "what is really needed is an argument to the effect that the wavefunction does not have such potency as the Everettians attribute to it", with "potency" meaning the WF being able "on its own to account for physical systems, apparatuses, people, etc.", in other words, to play the role of a substance bearing contingent properties: if the WF alone is substance enough to do all that, the introduction of a further substance in the form of particles indeed seems unnecessary.

It can certainly be denied that the WF *must* be regarded as a substance, or a substance-like entity. Yet, if it is regarded as such, strong arguments are required for maintaining that the particle configuration adds necessary (read: non-superfluous) structure to the theory. Irrespective of that, TBM avoids this objection from the outset. With its particular interpretation of the WF, it doesn't share the Bohmian WF's risk of being interpreted as a substance or substance-like: ontologically, the WF and the particles in TBM are *not* on a par. Only the latter play the role of a substance, while the former is regarded as encoding a *dispositional property* of the universe to actualize certain configurations. (Alternatively, adopting a nomological interpretation as discussed above, avoids the problem even more clearly.) Hence, concerns about substance dualism that apply to BM, as discussed by Valentini and Brown, do not carry over to TBM. More to the point, not being a substance itself, the WF alone cannot contain many worlds, or anything, for that matter: unless assuming a realist stance regarding properties without bearers, the WF in TBM is not enough to account for appearances including physical systems, people, etc., given material substance of some sort is required for that. Hence, TBM is not many worlds in denial.

In short: if material substance is required, then either the WF itself is substantial, in which case the configuration is superfluous, or the WF is not substantial, in which case the configuration is not superfluous. Since for TBM, we do require material substance, and do interpret the WF as not being substantial, the configuration is not superfluous. So the objection does not carry over to TBM. The upshot of all these considerations is that TBM evades the risk of having its basis for the definition of worlds undermined. The fear that this might be the case is unfounded.

4.1.2 Psychophysical Parallelism

As has just been argued, the substance or substance-like entity in TBM's ontology accounting for appearances is not to be found in the WF, but in the particle configuration. Whether one adopts this view, ultimately, is a matter of interpretive choice: one can instead choose to think of the WF as a substance, thereby rendering the configuration superfluous. Then, one should consider adopting the Everettian view. Both options are viable. But in any case, this decision has some import on how to define worlds within the resulting framework: Everettians must construe worlds from the WF. There is nothing else. In TBM, we are best advised to construe worlds from configurations, for they are the substance. And further down the line, deciding to take either the WF or the configuration as the basis on which to define worlds amounts to deciding for a stance on psychophysical parallelism (PPP): one must decide which part of the ontology of a theory shall be regarded as corresponding to an entity that can figure as an observable and that represents the physical that parallels the psychological in psychophysical parallelism. As Barbour (1999, 226) puts it: "Any scientific theory must establish a postulate of *psychophysical parallelism*:

it is necessary to say what elements of the physical theory correspond to actual conscious experience".

In the Everett interpretation, as Barbour points out, the individual branches of the WF fulfil that role. In TBM, on the other hand, worlds defined over configurations fulfil it. Barbour (1999, 302) and Bell agree with this choice: "Now, what did Bell regard as the physical counterpart of psychological experience? Is it in the wave function, as Everett and many others have assumed, or in matter configurations? Bell, like myself, opts for the latter: 'It is ... from the x s [(the configurations)], rather than from ψ, that in this theory we suppose 'observables' to be constructed. It is in terms of the x s that we would define a 'psycho-physical parallelism'—if we were pressed to go so far.' [(Bell, 1987a, 128)]".

Let's dwell a bit on that notion of psychophysical parallelism (PPP). As Heidelberger (2003) shows, PPP can be understood in several ways, ranging from a mere correlation between the physical and the mental that bears no metaphysical weight and allows for a strict substance dualism, to the idea that everything physical has some mental aspect to it (panpsychism). In between lies the idea that the mental is an *aspect* of *some* physical things, like, e.g., brains (Heidelberger, 2003, 238): "The mental and the physical are [...] two different aspects of one and the same entity" (double aspect PPP).

Barbour doesn't really say where exactly in this spectrum he locates his own account of psychophysical parallelism. On the one hand, he speaks of *correspondence* between elements of the physical theory and conscious experience, which suggests a rather weak form of PPP. On the other hand, it seems clear that he has a specific physical structure in mind, i.e., the brain, the different states of which correspond to experience, thereby suggesting a somewhat stronger PPP (Barbour, 1999, 26.): "However, there is increasing evidence that certain mental states and activities are correlated with certain physical states in different specific regions of the brain. This makes it natural to assume, as was done long ago, that there is psychophysical parallelism: conscious states somehow reflect physical states in the brain." Also (ibid., 266): "It is well known that much processing goes on in the brain and, employing normal temporal language, we can confidently assert that what we seem to experience in one instant is the product of the processing of data [...]."

I will come back later to this issue when discussing its bearings on the construction not only of worlds, but also of world histories; if one adopts a form of PPP that makes claims stronger than mere correlation between the physical and the mental, and if one considers the findings of modern neuroscience, an account that conceives of world histories in terms of processes of change between configurations rather than individual configurations alone suggests itself.

But before further discussing PPP and its import on the construal of world histories, we must tackle the question what a world in TBM is supposed to be. After all, world histories are supposed to consist of instantiations of worlds, standing in an appropriate relation to each other. What that relation is will be spelled out later. First, we must account for the relata featuring in said relation. As has been argued, worlds, in the framework of TBM, should best be construed in terms of particle configurations. However, they must not be identified with individual configurations simpliciter. The following paragraph will develop this approach.

4.2 What Is a World?

As mentioned in the introduction to Tychistic Bohmian Mechanics, TBM stems from Bell's Everett (?) theory. However, comparing them closely reveals that the two are far from being the same theory, for they differ in an important aspect—their construal of worlds. The just mentioned identification of worlds with individual configurations simpliciter is what Bell had in mind when developing—and immediately rejecting—his Everett (?) theory (Bell, 1987a, 133): "Thus instantaneous classical configurations x are supposed to exist, and to be distributed in the comparison class of possible worlds with probability $|\psi|^2$". Note that Bell uses the terms "configuration" and "world" interchangeably: possible worlds—not to be confused with possible worlds as in Lewis' (1986) modal realism—are simply configurations currently not actualized. Actualized configurations, accordingly, are the same as instantiated worlds. In short: for Bell, individual, instantaneous configurations *are* worlds.[6]

We would be ill-advised to simply adopt Bell's notion of a world, for it is this identification of worlds and individual configurations that leads him to his objection of temporal solipsism, and, a fortiori, to his rejection of the Everett (?) theory. We must be careful that TBM does not fall prey to the same objection. The problem of temporal solipsism has been introduced earlier. It is the reason for Dürr's and my construal of worlds, or rather: world histories, in terms of world branches. To summarize the issue: once worlds are identified with individual configurations as in Bell's account, as soon as the configuration changes, a different world is instantiated. And with this change between individual configurations being discontinuous, worlds are isolated from each other, not only in configuration space, but also in time. On this construal, no connection between worlds can be had. Reference to events future and past—from within a given world—is empty. Worlds, then, have no history in any meaningful sense of the word.

The solution to this problem presented in Dürr and Ehmann (2021) probably isn't quite satisfactory, in that it lacks a detailed description of how our experience of temporal history emerges from instantaneous configurations: our world branch is actualized infinitely many times in each finite time interval, but how exactly the individual actualisations are connected, other than by being in the support of the same world branch, hasn't been made clear yet. Besides, relying too heavily on the

[6] See also his usage of the term at other places in his 1987a: "Let this 'world' be simply a large number N of particles" (127); "the initial configuration x is chosen at random from an ensemble of configurations" (129); "a comparison class of possible initial worlds" (129); "a hypothetical ensemble (of worlds!)" (129).

branching structure of the WF for the definition of worlds defies the purpose of adopting a Bohmian particle ontology, and doesn't chime with an interpretation of the WF as encoding a dispositional property.

Time capsules, on the other hand, do provide an explanation of our temporal experience. But it can be contested that this explanation is ideal, especially if we aren't forced to assume that there is no external, universal time. Barbour's time-capsule approach will be discussed in greater detail shortly. For now, it shall suffice to say that we want to construe worlds differently.

In order to avoid Bell's criticism, our notion of worlds has to be able to do more than worlds in Bell's sense, identified with individual configurations, can accomplish. At the same time, we must not resort to time capsules or Everettian world branches alone. As hinted at above, in the end, we want to be able to recover the (quasi-classical) manifest image from TBM's scientific image. As Bell (1987a, 125 f.) points out, "classical concepts have not [...] been expelled from physics. On the contrary, they remain essential on the 'macroscopic' scale, for '... it is decisive to recognize that, however far the phenomena transcend the scope of classical physical explanation, the account of all evidence must be expressed in classical terms' [(Bohr, 1949)]. Thus contemporary theory employs both quantum wave functions ψ and classical variables x, and a description of any sufficiently large part of the world[7] involves both". Accordingly, I don't deem it enough for TBM to give an account of the microscopic goings-on. An explanation must be given as to how, at least in principle, macroscopic appearances, including the phenomenon of continuous macroscopic change over time, can emerge from the latter. In other words, we must "save the phenomena" by providing an account of psychophysical parallelism for TBM. Since we do not rely on Barbour's time capsules, individual configurations alone aren't a suitable choice for this purpose.

Instead, we have to construe worlds in a way that allows macroscopic objects to (quasi-)persist over change of configurations, but which also leaves room for their change based on the change of configurations. What we need is a notion of worlds that is able to macroscopically group together configurations that are different microscopically, such that worlds of finite, non-zero volume in configuration space can be construed. In the previous chapter, we have developed such a notion of worlds for the framework of BSM. In what follows, we shall apply it to the case of TBM.

Just for the sake of argument, let's *preliminarily* define a world in TBM as a collection of configurations generated from "coarse-graining" over possible configurations such that the resulting collection ($\text{WORLD}_{\text{TBM}}$) consists of configurations that are macroscopically *identical*, i.e., that instantiate the *same* macro state of the universe.

[7] Presumably, this instance of the term "world" does not refer to a single configuration, but the universe in total.

Already this preliminary definition avoids the aforementioned problem of Bell's identification of worlds and individual configurations. However, we won't stick with it for long, since it allows for partitioning of configuration space into *disjoint* sets, similar to the partitioning of phase space into *disjoint* macro regions in the usual framework of BSM. Recall that a macro region Γ_{M_i} in standard BSM is a set of micro states $x \in \Gamma$ all instantiating the same macro state M_i, and only M_i. That is to say: Macro regions in standard BSM do not overlap. For reasons extensively discussed earlier, this notion of a macro region as an equivalence class in phase space has been replaced by a definition of "world" for BSM: a world $\mathcal{W}_i$ is the set of all x_j that satisfy SS (i.e., are sufficiently similar in the sense of instantiating indistinguishable macro states) with respect to a given x_i:

$$\forall x_i \exists \mathcal{W}_i \left(\forall x_j \left(x_j \in \mathcal{W}_i \leftrightarrow x_j SS x_i \right) \right)$$

(*world membership*$_{\text{BSM}}$)

According to the new definition *world membership*$_{\text{BSM}}$, *all* micro states in a world are required to be sufficiently similar *not to all others*, but *to the central micro state* x_i around which the world $\mathcal{W}_i$ is construed. From this weaker requirement, it follows that not necessarily all $x_j \in \mathcal{W}_i$ are sufficiently similar to each other: SS is a nontransitive relation, and only requiring that all $x_j \in \mathcal{W}_i$ fulfil SS with respect to the central x_i leaves open the possibility that two other micro states $x_j, x_k \in \mathcal{W}_i$ do not stand in relation SS to each other. At the same time, a micro state x_j being sufficiently similar to a given central micro state x_i and hence belonging to $\mathcal{W}_i$ around x_i may also be sufficiently similar to another micro state, x_h, say, and hence belong to the world $\mathcal{W}_h$ around x_h.

This new definition of worlds in the case of BSM allowed for overlapping worlds, and hence solved the issue of jumping macro values despite a continuous micro dynamics and in light of the demand for reducibility of thermodynamics to BSM. In TBM, we don't have such a continuous micro dynamics. But we still need to recover (FAPP-)continuous macroscopic change. I.e., we need to provide an account of (FAPP-)continuous transitions between different macro states, similar as in BSM. Accordingly, adopting the approach we had developed for BSM and applying it to the case of TBM promises to be fruitful.

Thus, we reuse the definition of worlds we developed for BSM and adapt it for the TBM framework. Some slight modifications are necessary, though. In TBM, we are not dealing with micro states x in $6N$-dimensional phase space Γ (or, for systems of constant energy, in the $(6N-1)$-dimensional energy hypersurface Γ_E), but with configurations Q in $3N$-dimensional configuration space $\mathbb{R}^{3N}$.

Accordingly, a world $\mathcal{W}_i$ in TBM shall be defined as the set of all Q_j that satisfy SS (i.e., are sufficiently similar in the sense of instantiating indistinguishable[8] macro states) with respect to a given Q_i:[9]

$$\forall Q_i \exists \mathcal{W}_i \left(\forall Q_j \left(Q_j \in \mathcal{W}_i \leftrightarrow Q_j SS Q_i \right) \right)$$

(*world membership*$_{\text{TBM}}$)

Irrespective of the difference regarding the fundamental space within which the micro dynamics unfolds, and irrespective of the difference in the micro dynamics itself, the feature that worlds defined via SS may overlap, and everything this brings with it, carries over from the case of BSM to the case of TBM.[10]

Despite worlds possibly overlapping, *world membership*$_{\text{TBM}}$ guarantees—via sufficient similarity, and in accordance with our usual experiences—that macro states instantiated by configurations belonging to a world don't contradict the macro state instantiated by its central configuration, e.g., by there being dinosaurs in Hong Kong in one, but not in the other. In this sense it, suppresses inconsistent macro states: the respective configurations, differing to an extent that they contradict each other on the macroscopic level, certainly are macroscopically distinguishable. Hence, SS does not put them in each other's worlds. Consequentially, the application of SS leads to individuation (separation, "branching") of worlds whenever the grouping of configurations into a world would lead to inconsistencies with its central configuration.[11] But note that the requirement of sufficient similarity does not suppress the grouping of microscopically inconsistent configurations into one world: such configurations, differing only microscopically while being

[8] As opposed to what has been said above, here, being "indistinguishable" is not necessarily to be understood as being indistinguishable for some external observer or measuring device. The configuration Q usually refers to the configuration of the entire universe, understood as everything there is. By definition, there is no observer or device external to it. In particular, speaking of (in)distinguishability in this context does not imply the existence of a mind for which states are distinguishable or not. Rather, as shall be spelled out in some detail later, saying that macro states are indistinguishable simply means that these states do not amount to a macroscopic difference for a given macroscopic object (MO) within a configuration. Likewise, saying that macro states are distinguishable means that these states do amount to macroscopic difference for the MO. Such an MO *can* be a mind-bearing observer, but it *doesn't have to* be one.

[9] Despite being suitable for our present purposes, this definition is still somewhat preliminary, for the following reason: since our definition of worlds via SS hinges on the notion of macroscopic perspective, the meaning of the latter must be made explicit. As we will see, the distinction between the micro- and the macroscopic itself is relative—it differs, depending on the system or object in question. This, we need to account for in our definition. Likewise, we must be more specific regarding the overlap of worlds. Other than for the case of BSM, in order to construe world histories in TBM, a special condition on how worlds overlap—*overlap proper*—must be met.

[10] Recall from above that SS is reflexive, symmetric and nontransitive.

[11] Cf. Barbour's (1994, 408) similar requirement for records contained in time capsules: "A time capsule […] appears to contain records of the past, and these records are mutually consistent".

macroscopically indistinguishable from a central configuration, still belong to the world of the latter. This is to say that *world membership*$_{\text{TBM}}$ also guarantees that individuation of worlds in this sense only occurs when necessary. Preventing such "over-individuation", SS avoids the problematic identification of configurations and worlds: with a microscopic difference being sufficient for the individuation of worlds, every world would contain only one configuration, since every configuration differs from every other configuration microscopically.[12] Accordingly, worlds could not overlap, which ultimately leads to temporal solipsism. For example, on such an account, it is utterly unclear how the setup of an experiment and the recording of its results hang together (unless we assume that all of this emerges as an illusion from within one Barbourian time capsule). It is by avoiding over-individuation that TBM evades temporal solipsism.

4.3 Superpositions in TBM

One should be careful, though, to not confuse the grouping of microscopically inconsistent configurations with a superposition. Superpositions are a feature of the universal WF and its dynamics. It evolves independently of the actualized particle configurations, and likewise of their grouping into worlds. If it contains superpositions, it does so irrespective of how configuration space is coarse-grained into worlds, and which macro states are instantiated.

Having said this, it might be helpful to see how TBM accounts for interference experiments probing superpositions by looking at an example. This might also be useful to render the similarities and differences between TBM and some of its main competitors a bit clearer. Recall from the introduction to TBM my presentation of the measurement problem, following Maudlin (1995). One can categorise different quantum theories and/or interpretations thereof based on which route they take for the solution of this problem in dealing with the general presence of superpositions. In the following, before coming to TBM, I present one example from each of these categories: so-called orthodox quantum mechanics, often dubbed the Copenhagen interpretation, not accepting the unaltered Schrödinger dynamics and modifying it by adding a collapse postulate; Many-Worlds (or Everett) theory, replacing (universal) value indefiniteness by value multiplicity (cf. Wallace, 2012) for "macroscopic superpositions"; and BM, denying completeness of the WF, adding Bohmian particles/configurations to complete the state description. From this categorisation, we recover TBM's heritage: it is clearly to be grouped with BM, in that it adds particles/configurations to complete the state, rendering the complete state value definite at any given time while accepting the unaltered Schrödinger dynamics. Yet, despite being value definite at any given time, TBM's configuration is not determined by configurations at other times, for it lacks the deterministic particle dynamics, i.e.,

[12] This requires some qualifications, which will be provided below.

the GE, of standard BM. Maudlin (ibid., 13) refers to this as the "problem of effect". TBM solves it by appeal to worlds and world histories. Having to make this appeal is the price to pay for the excision of the deterministic GE. It results in the fact that, while still being value definite at any given time, different world histories, carving out the branching structure of the universal WF, are instantiated at random, rendering the theory many-worldish to the extent that none of the branches of the universal WF shall be forgotten—all of them are actualized again and again, although not simultaneously. In this restricted sense, TBM inherits value multiplicity from Everett.[13]

Take the usual single particle/wave packet double-slit experiment: A particle/wave packet source that can be set up such that individual particles/wave packets are emitted, one at a time; a double slit the particle/wave packet is passing through; and a screen for measuring the particle/wave packet position behind the double slit. Despite the system generally being in a superposition, we do not observe a macroscopic superposition at the screen.[14] Now, different theories/interpretations tell different stories of what is going on when this experiment is carried out.

According to Copenhagen, the wave packet, travelling through the apparatus, is in a superposition, until, upon measurement, it collapses onto a definite position state, with probabilities according to BR. What counts as a measurement, and accordingly, where and when collapse occurs, depends on how we define and divide the quantum system to be investigated from its environment, construed as classical. An experimenter in the lab, e.g., would place the measuring device, in this case the screen, on the opposite side of this so-called Heisenberg cut than the system to be measured. A colleague of the experimenter, however, might place the cut after the first experimenter, thereby treating the latter as part of the system under investigation, and so on.[15] This is of importance for our considerations: placing the Heisenberg cut decides on where and when the collapse of the WF happens. And it is the collapse of the WF that is supposed to account for the quantum-to-classical transition in Copenhagen, in the sense that it accounts for the fact that we do not observe a macroscopic superposition in single events.

By contrast to Copenhagen, Everett does not resort to a collapse postulate in order to explain the absence of macroscopic superpositions from our observations. Rather, decoherence accounts for the vanishing of interference terms, resulting in the branching structure of the WF, i.e., FAPP-independent, quasi-classical world branches all of which are realized simultaneously. As is regularly stated,

[13] In the less restricted sense as used by the Everett community, TBM does not feature value multiplicity: at each instant, only one configuration is actual, thus only one branch or world history actualized. Thus, at each instant, there is no multiplicity of values in the Everettian sense. Rather, multiplicity in the restricted sense appropriate for TBM appears only when temporally coarse graining over finite intervals of time.

[14] In a single event, that is. Of course, we are "observing" macroscopic superposition, in the sense that in a sufficiently large ensemble of single events, events approximately are distributed with Born statistics over the screen. That is, after many individual runs of the experiment, an interference pattern emerges on the screen.

[15] Cf. e.g., Schlosshauer and Camilleri (2011) and references therein.

decoherence alone provides no solution to the measurement problem, for it does not do away with value indefiniteness. Instead, Everett's solution to the measurement problem is the replacement of value indefiniteness by value multiplicity for macroscopic superpositions. Universally, superpositions remain. But as inhabitants of one such emergent world, we do not encounter macroscopic interference, despite superpositions still being present in the universal WF. As Wallace (2012, 36f.) puts it: "this, in a nutshell, is what the Everett interpretation claims about macroscopic quantum superpositions: they are just states of the world in which more than one macroscopically definite thing is happening at once. Macroscopic superpositions do not describe indefiniteness, they describe multiplicity". And later (ibid., 38): "If the correct way to understand such superpositions is as some sort of multiplicity, then our failure to observe that multiplicity is explained quite simply by the fact that we live in one of the 'worlds' and the other ones don't interact with ours strongly enough for us to detect them." Note that branching via decoherence depends on "magnifying" microscopic superpositions (ibid., 99): "branching is caused by any process which magnifies microscopic superpositions up to the level where decoherence kicks in". Everettian worlds, thus, are an emergent feature of a higher, macroscopic level of description, just like worlds in TBM are. Here, too, the distinction between the micro- and the macroscopic is rather vague (ibid, 100): "there is no 'finest' choice of branching structure: as we fine-grain our decoherent history space, we will eventually reach a point where interference between branches ceases to be negligible, but there is no precise point where this occurs".

Standard BM also doesn't resort to WF collapse, although the notion of effective collapse is common. This does not refer to a collapse of the universal WF as in Copenhagen, however. It basically means ignoring those branches of the WF that, after decoherence, are not effective in guiding the particle on its trajectory, for the interference term being vanishingly small. That branch of the WF in the support of which the particle resides is then called the effective WF. But again, the universal WF does not collapse in BM. As described in Dürr and Teufel (2009, 180), "this collapse is not a physical process, but an act of convenience. It is introduced because it would simply be uneconomical to keep the ineffective wave functions, and the price we pay for forgetting them amounts to nothing". Here, too, the superposition in the universal WF remains, but the particle selects one branch that is realized. Which branch that is—unbeknownst to the experimenter, according to the usual reading of the QEH—is already determined at the preparation of the experiment, i.e., in the particle source. Over the course of many repetitions, slight deviations in the preparation correspond to different branches being effective, such that different trajectories are realized, resulting in an interference pattern to appear on the screen.

In TBM, things are a bit different. The QEH, here, does not provide a measure of ignorance about the precise preparation of the particle. Rather, it provides a stochastic guiding law for the actualization of configurations. As we have seen above, SS groups different configurations, amounting to a microscopic superposition but being sufficiently similar to a central configuration macroscopically, into the same world, as long as the superposition has not been blown up to macroscopicity by resulting in a macroscopic effect. Decoherence in TBM results in branching of the WF, just as it does in Everett and standard BM. But neither are these branches all realized simultaneously, nor are any of them effectively discarded. Let's denote the configuration with the particle residing at a position within the left slit Q_l, and the configuration with the particle residing at a position within the right slit Q_r. These do not yet correspond to different branches being actualized/distinguishable macro states being instantiated, because they are macroscopically indistinguishable: via SS, $Q_r \in \mathcal{W}_l$ and $Q_l \in \mathcal{W}_r$. Compare the presentation of the Everettian account of this situation above and in particular, note the similar appeal to the microscopic-macroscopic-distinction: the quantum state is not yet decohered. No magnification of the microscopic superposition has taken place so far. Hence, although the notions are vague, it is clear that, in the Everettian terminology, branching has not yet occurred. Likewise, in TBM, worlds have not yet been individuated, i.e., configurations still lie in each other's worlds: with the particles residing at different positions, but both configurations being sufficiently similar to each other, their worlds "contain" a microscopic superposition. That is, these worlds overlap:[16] via SS, $Q_r \in \mathcal{W}_l$ and $Q_l \in \mathcal{W}_r$. It is only at a later stage of the experiment, with different macroscopic points on the screen lighting up, that configurations $Q_{l'}$ and $Q_{r'}$ are macroscopically distinguishable.[17] Other than Q_l and Q_r, the former are not sufficiently similar, i.e., do not instantiate indistinguishable macro states. Thus, $Q_{r'} \notin \mathcal{W}_{l'}$ and $Q_{l'} \notin \mathcal{W}_{r'}$. It is at this point where, in the Everettian picture, branching takes place. We see that it is not the presence of a superposition per se, but the presence of a *macroscopic* inconsistency, that individuates worlds, both in Everett and TBM. In particular, microscopic superpositions alone have no bearing on the individuation of worlds. Worlds, and world histories, are thus an emergent feature, like branches in Everett. In fact, world histories "trace" Everettian branches. Just like branching in Everett (cf. Wallace, 2012, 3.11), the individuation of worlds depends on what we take to be sufficiently similar on the macroscopic level.

By now, I already have used the terms "microscopic" and "macroscopic" extensively without making particularly clear what they are supposed to mean. Let me make up for that in the following paragraph.

4.4 The Microscopic-Macroscopic-Distinction and Its Import on the Construal of Worlds

It should have become evident by now that the distinction between the micro- and macroscopic is not very precise. Worlds, construed via a relation of sufficient similarity, spelled out in terms of macroscopic indistinguishability, thus are quite

[16] As will become clear later, they not only overlap, but overlap *properly*.

[17] A reminder: on TBM, configurations at different times are stochastically independent. Hence, one must not think that $Q_{l'}$ somehow follows from Q_l. That is, it is not possible to infer from the position at which the screen lights up the position of the particle at an earlier time.

pragmatic and relative. They are pragmatic, first and foremost, because it often depends on practical limitations between which states we can distinguish, and between which we can't. In the chapter on BSM, I have elaborated that macroscopic (in)distinguishability is observer- or instrument-relative: whether two macro states can be told apart or not depends on "resolution power"—the abilities and limitations of our sensory apparatus, and/or on the precision of the instruments we use. In this sense, macro states are distinguishable (or not) relative$_{\text{observer}}$ to a given observer and/or instrument with a certain resolution power.

(In)distinguishability also depends on our point of view: from far away, two states might be indistinguishable that are distinguishable at close distance. Consider two pixels on a computer screen. If one is lighting up, the other is dark, and vice versa. Say that, very close to the screen, you are able to tell these states apart. But from a certain distance and beyond, you aren't. Your sensory apparatus doesn't change, in terms of its abilities (in this case, quite literally the resolution power of your eyes), depending on your distance from the screen. But the *apparent* separation between two points is smaller from afar than it is from close by. So at some distance, your resolution power doesn't suffice anymore to resolve the spatial distance between the (middle points) of the two pixels. So indeed, it seems that distinguishability does not only depend on resolution power, but also on other things, like perspective and point of view.

Think of everyday phenomena: how far must the moon have wandered over the night sky for you to recognize that it now is somewhere different from where it was before? How about a snail? The perceived angular velocity of the moon, roughly being the same as the true angular velocity of the earth's rotation around its own axis, is about 360°/day. Helix pomatia, at a distance of one metre, and with a speed of 4.2 metres/hour, can run 16 full circles around you in the time the moon completes one circle. Its perceived angular velocity, from your perspective, is about 16 times as high as the moon's.

With distinguishability not only depending on resolution power, but also on perspective, the construal of worlds, making use of distinguishability, also depends on perspective: configurations that, from your perspective, are distinguishable, and hence don't belong to each other's worlds, might not be distinguishable from the perspective of someone else, or, in general, from the perspective of another being, or system.[18] This other being might be a million light years away, or as big as the solar system, and hence not being affected by the difference between, for example, your cat sitting on the couch or sleeping on your bed, just like you are not affected

[18] As pointed out above, the notion of macroscopic (in)distinguishability is not reserved for sentient beings (consciously) observing differences between macro states. In a broader sense, adequate for our treatment of TBM, two states being macroscopically distinguishable simply means that it makes a macroscopic difference for a given macroscopic object (MO) which one is instantiated, and likewise, two states being macroscopically indistinguishable simply means that it makes no macroscopic difference for a given MO which one is instantiated. Such MOs may very well be sentient beings, but they don't have to be. Speaking of "(in)distinguishability" is a convenient façon de parler that should not be taken to imply that a (conscious) observer or, in general, a mind of some sort or other is of fundamental importance.

by the difference between two slightly different molecular compositions of the beast. It is here where the relativity of the terms "microscopic" and "macroscopic" enters the scene. It is a matter of perspective, not least of size, what does and does not count as micro- or macroscopic. Hence, the usual construal of the microscopic-macroscopic-distinction as two levels of description, appropriate for two different length scales defined in absolute terms, is too simple.

Let's elaborate on this. Length is a measure usually defined over the positive reals (see e.g., Tao, 2011), such that lengths, in general, can have any positive real-valued magnitude. The definition of a standard unit, which, for length, is the metre, is a pragmatic convention. And so is picking out a certain interval of the reals, measuring it in terms of some standard unit, and call it "the macroscopic length scale" or saying that it belongs to "the macroscopic realm". The same goes for the microscopic. Especially in descriptions like "the scale of objects that can no longer be seen by the naked eye" of the microscopic scale, the pragmatic nature of such assignments shows. See, for example, the definition in Reif (1965, 2): "It is useful to introduce a distinction between the sizes of systems whose description may be of interest. We shall call a system "*micro*scopic" (i.e., "*small* scale") if it is roughly of atomic dimensions or smaller (say of the order of 10 Å or less). For example, the system might be a molecule. On the other hand, we shall call a system "*macro*scopic" (i.e., "*large* scale") when it is large enough to be visible in the ordinary sense (say greater than 1 micron, so that it can at least be observed with a microscope using ordinary light)."

Definitions like this are not only pragmatic, but highly dependent on the contingencies of one part of the human sensory apparatus.[19] But there are other senses, and modes of sensation, in humans as well as in other beings with different sensory abilities. In general, different systems might only be affected by objects (or events, or processes) at greater scales or already be affected by objects (or events, or processes) at smaller scales than we as humans typically are. This suggests, instead of defining the micro- and macroscopic relative to the human sensory apparatus, that we define these notions relative to any two systems in general. After all, we are interested in describing systems and their relations to each other, whether humans are involved or not. And by doing so, we don't lose anything. What we hitherto have called microscopic, with respect to our sensory abilities, can still be called microscopic, as a result of the relation that holds between the system we call "us" and the system to be described. On the other hand, though, we gain the possibility of using the microscopic-macroscopic-distinction in other cases not involving humans or other beings of a certain size. What will matter are only relative measures and how things (objects/events/processes) at different scales affect each other.

One might be tempted, for TBM, to simply adopt a distinction between the micro- and the macroscopic based on decoherence:[20] a system, then, can be called

[19] Jaeger (2014, 897) also cites this definition and comments: "The appeal of such a notion is not surprising, given the dominance of the visual sense in humans".

[20] Cf. also the discussion in the next paragraph.

macroscopic once it contains sufficiently many degrees of freedom, or is large/heavy enough such that "the environment [...] continually interacts with, and becomes entangled with, that system" (Wallace, 2012, 77). Under such circumstances, systems decohere in a sufficiently short time interval, such that interference becomes practically irrelevant. However, such a construal does not evade vagueness. Recall (ibid., 100): "there is no precise point where this occurs". More importantly though, decoherence is a process that accounts for the branching structure of the WF. This is fine for Everett, but not enough for our purposes. With the particle configuration as TBM's referent and basis for its psychophysical parallelism, our micro-macro-distinction should be spelled out in terms of objects and their constituents as being part of the universal configuration.

Let us therefore adopt the following characterization of the microscopic-macroscopic-distinction: a scale is called microscopic with respect to another, called macroscopic, if change on the level of the former doesn't necessarily imply change on the level of the latter, i.e., if it is possible that, from the perspective of the macroscopic object, nothing has changed despite actual changes happening on the level of the microscopic.[21]

Through this characterization, it becomes clear that, although the distinction is dependent on the relation of levels, the relation itself is objective: Whether it makes any difference if your cat sits on the bed or on the sofa, is relative to the level from which the situation is described. From your perspective, it does make a difference. Hence, the difference in the cat's position, described from the level of your perspective, is regarded as macroscopically distinguishable. Changing perspective to a relatively large scale can change that status. From the perspective of the solar system and the level used to describe it, the change in your cat's position is irrelevant, hence regarded as microscopic. Yet, despite this relative dependency on the level of description, it is objectively so that, from your perspective, the position of your cat is relevant, while it is not from the perspective of the solar system.

4.4.1 *The Microscopic and the Macroscopic Domain vs. the Quantum and the Classical Domain*

A characterization like the one just suggested can also help to prevent the conflation of the "quantum scale" or "quantum domain" with the microscopic, and of the "classical scale" or "classical domain" with the macroscopic, respectively. Not everything small is quantum, not everything big classical. Rather, we should use the terms quantum and classical for phenomena best described by the respective

[21] Note the resemblance of the micro-macro-distinction adopted for TBM, and for BSM (cf. the previous chapter), respectively. This is a recurring theme, arguably rendering TBM—as well as other particle-based quantum theories adopting the notion of worlds as collections of sufficiently similar micro configurations—especially suitable for the replacement of BSM in cases where a quantum rather than classical treatment is appropriate (see the following paragraph).

theories, independently of the size of the objects involved. Jaeger (2014, 896f.) discusses this topic (and traces the history of the use of the term macroscopic): "Related to the popularity of this approach [(i.e., the Copenhagen interpretation)] is the widespread assumption, which is often made without circumspection and most often made in popular discussions, that the macroscopic realm and the regime of classical physics are one and the same, sometimes resulting in the use of the terms classical and macroscopic interchangeably. This has occasionally been noted, for example, by Żurek in his well known discussion of the attempt to understand the transition from quantum-mechanical to classical-mechanical behavior through the notion of decoherence[:] 'In the absence of a crisp criterion to distinguish between quantum and classical, an identification of the classical with the macroscopic has often been tentatively accepted. The inadequacy of this approach has become apparent as a result of recent developments: A cryogenic version of the Weber bar—a gravity-wave detector—must be treated as a quantum harmonic oscillator even though it may weigh a ton…' [(Zurek, 2002)]" Later, Jaeger (2014, 898) goes on: "To reach the identification of the macroscopic with the classical, one would need a system being macroscopic as both necessary and sufficient for its being classical. As noted in the introduction, however, some quite massive gravity-wave detectors are required to be treated quantum mechanically. Moreover, a system's having large quantum numbers, its having large mass, and its having a large spatial extent do not always coincide: one example of this would be that of a compact astronomical object which is of "microscopic" spatial extent while also being considered "macroscopic" by virtue of a large mass—having a Schwarzschild radius of a size larger than an atomic nucleus but smaller than an atom."

These latter remarks suggest that what we have said above about length holds analogously for other measures. Defining intervals over the space of possible values and calling them the microscopic or macroscopic scale, first of all, is a matter of pragmatic convention, usually based on the human senses and everyday experiences, which are a contingent matter of fact. That doesn't change when considering other measures instead of length, e.g., mass. A definition based on relations between systems can reproduce the common notions of micro- and macroscopicity by relating the system of interest to systems roughly of the size, mass, or what have you we encounter in our everyday experience, which can be, if wished for, a human being. But it can also reproduce many more, for example by relating systems of the type just described to other, much bigger (or heavier) systems, a human being to the solar system, say. Suddenly, objects we encounter on a daily basis, including other human beings and ourselves, are rendered microscopic with respect to the macroscopic solar system.[22] As a consequence, it becomes almost trivial to explain why it would be wrong to conflate the domains thus variably rendered micro- or macroscopic with the domains of quantum or classical mechanics, despite the fact that *often* (not always!), the domain where quantum mechanics is adequately used for the

[22] Which seems to be an advantage. It can be regarded a theoretical virtue to exclude reference to contingencies like the usual size of humans from general descriptions, and another virtue to be able to reintroduce them as special cases when of interest.

description of phenomena is one of the small, light, etc., and the domain where classical mechanics is adequately used is one of the big, heavy, and so on, with "small", "light", "big" and "heavy" as judged from the human perspective.

So much for now about the microscopic-macroscopic-distinction. Earlier, I promised to come back to the issue of psychophysical parallelism. My claim was that an account of world histories in terms of processes of change between configurations will suggest itself, provided we adopt a more than minimal version of psychophysical parallelism and give some credence to the best neuroscientific theories we currently have. To first address these requirements, we shall make a short detour, discussing *a minor issue with Barbour's account.*

4.5 A Minor Issue with Barbour's Account

The issue I identify in Barbour's account is not a theoretical defect sensu stricto. Rather, what I find somewhat problematic is him ignoring, against what seems to be his own conviction, the important role that processes, and hence change—as opposed to states, individual events or static objects—play when it comes to psychological experience and its physiological basis,[23] and not providing a satisfying answer to the question of how, in his account, conscious experience arises from static configurations. Processes might not be metaphysically—let alone logically—necessary for experience, but modern neuroscience suggests that, at least for us, physiologically they are. Given we can trust our best theories, we should believe that experience arises from processes, not static states.[24] There's even a measurable minimum amount of time required—in our context, this translates to a minimum amount of change—in order for consciousness to emerge from an active group of neurons.[25] We should not sweep this evidence under the carpet light-heartedly, especially not if we don't have to.

Over the course of more than two decades, Barbour in his writings[26] develops a timeless cosmology, including an account of psychophysical parallelism. It bears some resemblance, but is far from identical, to the account presented in this book. As has been pointed out before, it is similar in that it features instantaneous configurations at the heart of its ontology, and as the physical basis of psychological phenomena: Barbour (1999, 302), following Bell, regards particle configurations as "the physical counterpart of psychological experience". However, contrasting my

[23] Cf. Barbour (1999, 266): "it is well known that much processing goes on in the brain and, employing normal temporal language, we can confidently assert that what we seem to experience in one instant is the product of the processing of data coming from a finite span of time".

[24] Despite the common usage of the term "brain state", especially in the philosophy of mind. Such "brain states" are not states in the strict sense of an individual, unchanging micro configuration of the brain.

[25] See, e.g., Sacks (2017, 173ff.)

[26] Cf. Barbour (1994, 1999, 2009, 2020).

account, he deems individual configurations—albeit somewhat special configurations, i.e., time capsules—sufficient to give rise to psychological experience. In this respect, Barbour's account resembles Bell's Everett (?) theory. What's more, he not only conceives of worlds as individual configurations (like Bell did), but also as eternal, static objects. This, I hold, gives rise to concern when it comes to psycho-physical parallelism, understood in a more than minimal sense.

Intuitively, one should think: experience, in particular experience of motion, requires change of configurations.[27] Change in the configuration of what I experience as moving—the Kingfisher in Barbour's example[28]—and, at least as importantly, change in myself that reflects the change of the object I observe and allows me to experience that change. Both, change in the object as well as change in the subject—me—require change in the configuration of the universe. So, experience, all the more experience of motion, should not occur within individual, static configurations. Yet, Barbour in his account construes experience as being had from within such configurations. So, given that experience indeed requires change of configurations, Barbour's account is misguided.

For our purposes, we can set aside the question of how exactly change in the experiencing subject is supposed to be linked up with change of an external object, such as to adequately reflect the latter's movement. Already the case of experience simpliciter is enough to convey the sense in which Barbour's account is problematic. Given there are subjects that have experiences of some sort. Since these subjects are part of the universe they live in, and experience seems to require change in that which has the experience, it seems to be the case that the configuration of the universe should change in order for experiencing subjects to arise. But Barbour's account doesn't allow for change of the configuration. This poses a problem Barbour tries to circumvent by denying at least one of the premises that (1) experience requires change within the experiencing subject, and (2) change within an experiencing subject requires change in the configuration of the universe. I doubt that he would object to the second premise: since the experiencing subject is part of the configuration of the universe, every change within the former amounts to change within the latter, if only on the microscopic level. Instead, he denies the intuitive premise (1) and introduces the notion of *time capsules*: individual, static configurations that contain what seem to be records of the past. "Somehow", by comparing the parts of the configuration that are regarded as records, the illusion of time flowing is said to arise. Let's take the existence of such time capsules for granted, if only for the sake of the argument. Does this solve the problem? I hold that it doesn't, even though at first glance, one might think that time capsules are sufficiently structured to account for experience. But the problem of *experience without change* is

[27] Note that this really is just an intuition. Intuitions can be wrong, so without further investigation, they never suffice to reject an account. Accordingly, it would be premature to reject Barbour's account just because it doesn't chime with our intuitions. But they can very well guide us in asking the questions that need answering. Personally, I deem it useful to let intuitions tell me where to look, but not what to find.

[28] Barbour (1999), passim.

merely pushed back a little. It can now be formulated as follows: given there are time capsules, how is the operation of comparing several records contained in a time capsule supposed to work from within such an individual, static configuration? Doesn't such an operation itself require a process to supervene on? Tailored to the case of human experience: how does a brain compare the records contained within a configuration while itself being caught in a static state, i.e., without itself undergoing change, i.e., without brain processes happening?

Recall that Barbour advocates psychophysical parallelism: experience should somehow be correlated with something physical. And indeed, he does name this something—time capsules. Accordingly, my objection to Barbour's account is not that he doesn't meet his own demand to "establish a postulate of psychophysical parallelism". He clearly does that, and from a purely metaphysical or logical perspective, there is no problem with his proposal. It could well be the case that this "something physical" with which experience corresponds is a single, static state: "Above" each time capsule, a mind might "hover" and hallucinate change in accordance with the records present in its associated time capsule. Consequentially, Barbour's account does work, under the assumption of such a minimal form of PPP, featuring nothing but this unexplained correlation between mind and configuration. This amounts to a dualist form of PPP: other than the thin correlation between configuration and mind, nothing accounts for them hanging together. Now, my worries are that this minimalist PPP is not supported by our current best understanding of how the brain gives rise to experiences, and that, on top of that, even Barbour himself might envisage a stronger account. But irrespective of that latter point, in light of our best theories, one should at the very least provide some good arguments in favour of such an assumption, especially when putting forward a theory that so strongly hinges on PPP. Barbour does not provide such arguments. To the contrary, his remarks about the relation between brain processes and psychological experience even suggest that his idea of how psychophysical parallelism works might be far from minimal (ibid., 266): "it is well known that much processing goes on in the brain and, employing normal temporal language, we can confidently assert that what we seem to experience in one instant is the product of the processing of data coming from a finite span of time".

Charitably assuming that such a *product* of data processing can be fully present and experienced in a single instant, the question remains how the *processing* bringing about this product is supposed to happen within an unchanging configuration. That would amount to a process that, qua definition of what a process is, is not a process. Barbour, by resorting to explanations as quoted, at least implicitly seems to acknowledge that an account featuring more than a minimal form of PPP must allow for change of configurations. His qualification "employing normal temporal language" does not resolve the issue. Understandably, he resorts to such formulations, indicating that his talk of processes is merely a façon de parler, not to be taken literally, for there are no literal processes within static configurations. But then, he should give an alternative account of how experiences arise, in particular, how the brain is able to compare and put into order the several snapshots it contains, without real processes happening. But he doesn't provide such an account. All we get (ibid.,

266 f.) is "the working conjecture [...] that when we think we see motion at some instant, the underlying reality is that our brain at that instant contains data corresponding to several different positions of the object perceived to be in motion. My brain contains, at any one instant, several 'snapshots' at once. The brain, through the way in which it presents data to consciousness, somehow 'plays the movie' for me."

This sounds rather suspicious. I have no objection towards the idea that the brain can at one instant contain several records or 'snapshots' of an external object, each representing the latter in a different state. What I find questionable, in addition to the qualms already discussed, is the following: the brain in this account is presenting data to consciousness/"plays the movie" to the experiencing subject. This, taken literally, again amounts to a mind-body dualism that is at least not supported, despite not logically being ruled out, by Barbour's more than minimal psychophysical parallelism as reconstructed above. Finally, all this is supposed to happen "somehow", which is a rather hand-waving explanation, if it is an explanation at all. All things considered, it remains utterly unclear how experiences are supposed to emerge from a static configuration.

The problem of Barbour's account seems to be condensable into the following argument: in order to account for the illusion of change, he requires an entity—a brain—capable of comparing snapshots from within a static configuration. At the same time, it stands to reason that this entity can only fulfil this operation if it itself undergoes change, i.e., if brain processes can happen. But in a single, static configuration, there are no brain processes, and thus no entity capable of comparing snapshots from within it. Hence, there is no account for the illusion of change. Barbour's account seems to block itself.

Granted, this argument works only if one doesn't assume a mind strictly independent of any physical supervenience basis. Such an entity could carry out the necessary operation of comparing records contained in time capsules. But Barbour obviously doesn't make this assumption, given his requirement for PPP with configurations as the basis. After all, assuming an independent mind would render the whole effort moot: richly structured time capsules as the physical aspect of PPP, the comparison of records from which the illusion of time emerges, all this is superfluous if we simply assume some entity capable of having experiences independently of the physical basis. Since Barbour deems it necessary to go through all this trouble, I, in turn, deem it safe to say that an entity experiencing independently of a physical basis is not what he envisages for his account. Following his own presentation, it is the time capsule itself, via its rich internal structure, and not some independent entity, that accounts for experience (ibid., 267, my emphasis): "A time capsule [...] is so highly structured that *it creates the impression of motion*." Accordingly, I still see no way out of the predicament Barbour's account is in. On the one hand, he doesn't allow for change of configurations, because all configurations are supposed to be actualized eternally. On the other, he arguably assumes a more than minimal form of PPP that is too demanding to work with an account that doesn't allow for change of configurations. Of course, we must consider the possibilities that (a) I have misinterpreted Barbour with respect to his stance on PPP and he in fact deems a minimal PPP with its dualist consequences acceptable; and that (b) I have read him correctly and he had a more than minimal PPP in mind, but would, in light of the considerations above, reconsider and adopt a minimal PPP. In both cases, I have no objection against his account. Alas, I would not adopt it for TBM, if only for the simple reason that for TBM, we don't have to buy into a minimal PPP and, a fortiori, dualism: TBM allows for change of configurations, so we may as well try to come up with an account featuring a more than minimal form of PPP, one that doesn't throw overboard our current understanding of how the physical and mental hang together.

In light of all that has been said so far, I suggest considering a theory in which psychophysical parallelism is stronger than a mere claim of correlation while being compatible with the underlying ontology and dynamics (and with our current best theories in neurology and cognitive science). To be precise, we should consider an account that allows for change of configurations, in the sense of different configurations being actualized at different times, such that experience can arise from processes of change rather than from a static entity like time capsules. This is not a rejection of Barbour's account *as a whole*: the universal WF (or its replacement) still could be time-independent. The relevant difference between Barbour's account of the emergence of conscious experience and mine is that I don't assume conscious experience without physical change. So let's see whether we can come up with an account in which experience stems from processes, and that is compatible with TBM's discontinuous actualizations of configurations. The gist of such an account has already been provided, in the form of our definition of worlds—via *world membership*$_{\text{TBM}}$—above. Now, as announced earlier, we need to expand on this definition, and then distinguish between overlap simpliciter and overlap proper. This will allow us to construe (FAPP-) continuous world histories from successions of individual worlds, instantiated by individual configurations.

4.6 The Macroscopic Object

The first problem that poses itself when developing a theory that construes experience as emerging from processes rather than static configurations can be stated like this: the universe has to know how to change configurations (and choose their order) by itself. We cannot rely on a mind to do so. Here is the reason: as suggested by our current neurological understanding, cognition, experience, etc., in general what we would call "mind", correspond to brain processes, rather than brain states *sensu stricto*, understood as single, instantaneous configurations. These brain processes, as a series of configurations/changes of configurations, occur independently of what emerges from them, in the sense that the "emergent" feature is not causally affecting its "basis".[29] What has to happen in the universe, all by itself and without a mind

[29] The reader may think of this as a form of the principle of causal closure of the physical, as it is often named (cf. Kim 1993), although in TBM, individual actualizations of configurations, being stochastically independent, are not causally connected. The principle in the context of TBM hence

making it so, is a series of actualized configurations that forms a consistent history without macroscopic "jumps", but that allows for macroscopic change over time. Consecutive actualizations of configurations within such a history must resemble each other so closely that, macroscopically, i.e., for a given macroscopic object within those configurations, the actualization of one after the other does not amount to macroscopic change, but actualizations further apart temporally may amount to macroscopic change. Only then (which is not to be understood as temporal order, but as a conditional conjunctive) can minds emerge, given the emergence of minds requires processes to happen. Following Heidelberger (2003, 236), psychophysical parallelism in the sense of a "double aspect theory", where "the mental and the physical are [understood as] two different aspects of one and the same entity" is incompatible with the idea that the mind causes the process to happen. Conversely, it is incompatible with the idea that the process causes the mind to be (ibid., 238 f.): "[Double aspect PPP] is defined as noncausal and therefore noninteractionist. But this noncausal interpretation [...] results from the definition of the psychical and the physical in terms of the perspective in which something is given. Viewing the physical as something that causes the mental, or vice versa, results from scrambling differing perspectives. [...] We can demonstrate that distinguishing perspectives is nothing mysterious by considering a bent coin. It would be ridiculous to say that a dent on the head's side causes a bulge on the tail's side. While both sides of the coin are intimately connected, their joint occurrence has nothing to do with causality".

We now have two options. Either we abandon double aspect PPP, or we show that, within our theoretical framework, change is possible without being caused by a mind. Frankly, the latter turns out to be rather trivial: within TBM, the configuration changes all the time. But change of the configuration simpliciter is not enough. It is less trivial to relate configurations to each other in such a manner that not only individual worlds, but entire, historically consistent world histories can emerge. This is what we will discuss now.

First, we should agree that "brains", as we usually conceive of them, are but one possible way for objects to embody minds within the framework of double aspect PPP. In principle, there could be many physical systems from which minds of some sort could emerge. We don't need to go into the details about this here. All we need is to agree that mind-bearing physical systems don't necessarily have to resemble brains like ours. To reflect this, I speak of brain-equivalent objects (BEO) rather than "brains".

What's more, BEOs aren't necessary for the construction of worlds and world histories from configurations and configurational changes. They are only relevant when the phenomenal experiences of mind-bearing beings (such as persons) and their emergence within the framework of TBM are under investigation. Abstracting from such special phenomena, all we need in the more general case are objects that

---

does not entail that there are causal connections between physical events on the fundamental level. It simply means that there are no causal connections between the mental and the physical, in particular, that there isn't anything non-physical causing actualizations and/or grouping them together into worlds, in accordance with the double aspect form of PPP we adopt.

are macroscopic with respect to the level of particles and their configurations—i.e., the fundamental, microscopic level of TBM's ontology—such that they do not necessarily undergo macroscopic change as soon as they undergo microscopic change, but may be affected by microscopic change such that they do undergo macroscopic change. Since BEOs are just a special type of macroscopic objects (MO), what we are going to say about the latter with respect to worlds and world histories carries over to the former. In addition, BEOs will be able to have *experiences*, while MOs, in general, will not, (unless they are BEOs).

Next, we should make the definition of a world a bit more precise. First, we need to be clear about what parts of the configuration of the universe are relevant to an MO. As we will see, it turns out that ultimately, this is only the part of the configuration that makes up the MO itself. Whatever happens outside the MO is completely irrelevant to the MO as long as it does not affect the MO, i.e., the part of the configuration that makes up the MO.

If that sounds strange at first, think of the prominent brain in a vat scenario.[30] Incidentally, the TBM framework allows for scenarios that lie far beyond the outlandishness of this thought experiment. A quick reminder: in the brain in a vat scenario, we imagine a brain that is hooked up to a device, let's say a supercomputer, that generates neuronal input to the brain such that the brain hallucinates having a body and strolling along a beach, for example, while it actually sits in a vat filled with a liquid that provides for all its physiological needs, including nutrients, etc. Neuronal output of the brain is fed back to the computer and integrated into the simulation, so that the brain doesn't have the impression of living a movie, but instead of actually being able to do things with its body, manipulate the environment, and so on.

In this scenario, some implicit assumptions are made that, strictly speaking, are unnecessary within the TBM framework.

Firstly, we can cut off the connection to the simulation device. Underlying the idea that such a device is required is the assumption that a brain needs some input that is causally connected to some external system in order to experience either something that is not absurd, or even to experience something at all. (Dreams might be an attempt to find a counterexample to the latter, although even when dreaming, we are not completely deprived of sensory input.) But within TBM, what the brain—or BEO, in general—experiences solely depends on either (a) the configurational state it is in at the instant it is actualized (Barbour) or (b) the succession of configurational states within the world history it lives in (my account), as we will see later. But in any case, the BEO needs no external machine in order to experience a walk at the beach. It just needs to be contained in an actualized (a) time capsule or (b) world history in which it just happens to have that experience. There is no causal connection to anything external necessary.

Secondly, as an extension of what we have said about the computer, we can get rid of the vat, too. Introducing a vat in which the brain receives everything it

---

[30] See Putnam (1981), 'Brains in a vat'.

physiologically needs to function properly is only required if we make the (implicit) assumption that a brain cannot exist on its own but must be embedded causally in an environment that first of all allows for its existence and functioning. Under this assumption, an environment too hot or cold, e.g., would already ruin the scenario, for the brain would not function well or even die.

Making this assumption is not necessary in TBM, though. Within each individual, static configuration, there is an essential causal independence between the configuration of the particles that make up the BEO (or MO) in question and the configuration of the particles that make up all the rest: Whatever exists elsewhere within the configuration containing the MO is irrelevant for the MO's configuration. Thus, in principle, it is irrelevant for the existence of the MO, and, over the course of many actualizations, irrelevant for its experience (if this MO is a BEO capable of experience). In principle, the MO can be the only thing that exists in the universe, and still undergo the configurational change it just happens to undergo. And if this MO is a BEO, it can, accordingly, be the only thing that exists in the universe, and still experience whatever it just happens to experience; it needs no energy, or nutrients, or neural input from the outside. Due to the stochastic micro dynamics of the particles that make up the universal particle configuration, any object composed of a subset of those particles is, in principle, dynamically independent of all the other particles and objects in that configuration. So we can imagine a BEO, e.g., a brain, floating in outer space while having the experience of strolling at the beach. According to the time-capsule approach (a), as soon as this happens the right way, i.e., with the configuration of the brain containing the appropriate snapshots, this brain would indeed have the required experience. Following the world-history approach (b), it would take a sequence of changes between configurations for the brain to have that experience, but once the strange thing has happened and the brain has materialized in outer space, there might as well be further actualizations, such that the respective world history emerges. This way or that, this scenario certainly is possible, both in Barbour's account as well as in TBM.

Speaking of world histories, I shall now take the next step towards their construction in order to flesh out the remarks above, in which I appeal to their existence. As has been noted, in order to account for world histories emerging from randomly actualized configurations, we must differentiate between two types of overlap between worlds: overlap simpliciter and overlap proper. Only the latter will prove to be sufficient for world histories to arise, due to the fact that overlap simpliciter does not guarantee a FAPP-continuous evolution on the emergent level. To render it clear just when worlds in TBM overlap properly, and when they don't, we first have to take a look at TBM's configuration spaces and their inhabitants (i.e., possible configurations), for worlds will be constructed from the latter within the former. The relevant considerations unavoidably will include some digressions into particle metaphysics—particle identity and impenetrability in particular—especially in light of the desideratum of particle number conservation in non-relativistic quantum theories, such as TBM.

4.7 Configurations and Configuration Spaces

For a universe of N particles with d degrees of freedom, we can construct a dN-dimensional configuration space $\mathbb{R}^{dN}$. For physical space with 3 degrees of freedom, this amounts to a $3N$-dimensional configuration space $\mathbb{R}^{3N}$.

Let's assume for the moment some sort of principium identitatis indiscernibilium (PII) for physical particles, stating that particles are regarded as identical when they are indistinguishable, and as non-identical when they are distinguishable. For particles to be distinguishable, they need some kind of feature with respect to which they differ from each other. So, for example, an ontology where all particles are featureless, or an ontology where particles aren't featureless, but all instantiate the same features, such that they do not differ despite, must construe them as identical with each other. (For the time being, we are ignoring their spatiotemporal trajectories, which in some cases might be used to individuate them despite their featurelessness, or sameness with respect to their features. We will discuss this later.) Contrariwise, for an ontology where all particles are individuated, for example by their intrinsic properties, haecceity, or a little label (a rigid designator) that is attached to them, there are no identical particles except for the rather trivial case of self-identity. In between these examples lies the case where particles neither are equipped with primitive thisness or the like, nor are all featureless (or all have the same features), but have features—e.g., mass, electrical charge, etc.—in which they differ. Via these features, types of particles can be distinguished, e.g., electrons and positrons, but unless every particle is equipped with a unique combination of features, or with haecceity, this does not allow for the individuation of all particles there are. Two electrons, say, that do not differ in any of their features, cannot be individuated, although they are numerically different, and even reside at different positions in physical 3-space: just imagine a two-particle universe, with the two particles having identical features. If a demon had exchanged the particles a number of times, putting each at the position of the other and vice versa, without us knowing how often it did that, we would not be able to tell whether the particles in the end are at their original positions or had their positions swapped. There is nothing that could make any difference between the two possibilities. For the time being, this is what we mean by saying that the particles cannot be individuated.

Depending on the system in question and the particles involved, our representation of their configurations in configuration space may differ. Let's start with the simple case where all particles can be individuated due to them each possessing a unique combination of properties, a label, or what have you. In this case, permutations of particles always amount to a different configuration. Hence, all of configuration space is relevant. For an illustration, we assume a toy universe with two individuated particles and one spatial degree of freedom. In this toy universe, the complete two-dimensional plane of our configuration space $\mathbb{R}^2$ is relevant. Since particles at spatial positions q_1 and q_2 are individuals qua their discerning features, configurations $Q(q_1, q_2)$ and $Q'(q_1, q_2)$ (where particles have changed places, i.e., have exchanged their position values, such that q_2 takes on the value of q_1 and vice versa) are discernible as well, hence distinct. They amount to different facts. Consider the following example (Fig. 4.1): One particle resides at $q_1 = -1.5$ and another at $q_2 = 1$. In configuration space, this configuration is represented as the point $Q(-1.5,1)$ (the blue dot). Now let them change places, such that $q_1 = 1$ and $q_2 = -1.5$, represented by $Q'(1,-1.5)$ (the orange dot).

But what if we are dealing with indiscernible and hence identical particles at q_1 and q_2? Q and Q' then should amount to the same configuration, since it doesn't make any difference whether one is at -1.5 and the other at 1 or vice versa. That is to say, mere permutations of identical particles do not individuate configurations. Hence, when representing configurations of identical particles, our configuration space is structurally richer than the possible particle configurations it represents.

Arguably, given identical particles, one should use reduced configuration space to represent meaningfully different possible configurations. To make the formal representation and what it represents match again, we reduce the full configuration space such that the superfluous structure is ignored. For an illustrative example (Fig. 4.2), we can do this by simply mapping every configuration below the symmetry axis (red) to the other side.

For higher-dimensional configuration spaces, we can do similar rearrangements. If we expand our toy universe by adding a third particle, say, such that all three particles are identical, then there are $3! = 6$ possible permutations that amount to the same configuration. (In general, there are $N!$ permutations for N identical particles, all amounting to the same configuration.) In order to fix that, we get rid of all these permutations except one.

If we do this to all *possible* configurations, i.e., all points in configuration space, we arrive at the reduced configuration space representing all possible configurations

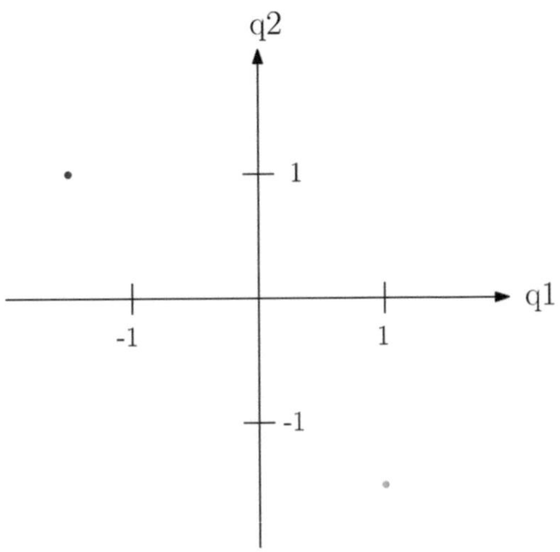

Fig. 4.1 Configuration space $\mathbb{R}^2$ of two discernible particles with one spatial degree of freedom. The orange and blue dots represent two distinct configurations, despite the fact that in both configurations, the same two spatial locations are occupied

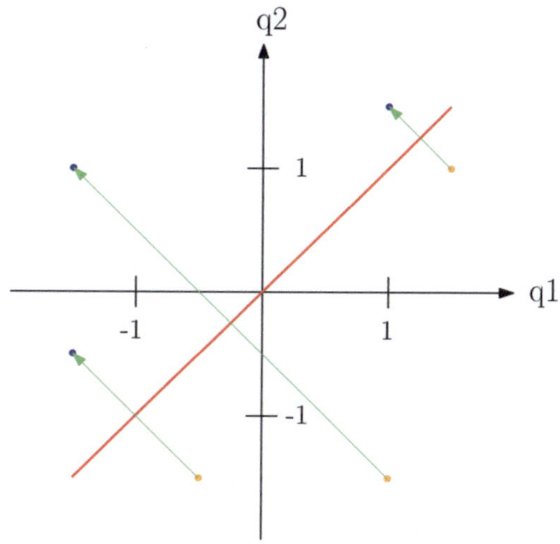

Fig. 4.2 Identical configurations are mapped onto each other

of a given system of identical particles that differ meaningfully, i.e., are not mere permutations (Brown et al., 1999, 230): "The 'reduced' space $\mathbb{R}^{Nd} / S_N$ […] is the quotient of $\mathbb{R}^{Nd}$ obtained by the action of the symmetric group S_N (the group of permutations P above)", the latter simply being defined (ibid., 229) as "an arbitrary permutation on the set $\{1,2,\ldots, N\}$" with N being the number of particles involved.

4.7.1 *Particle Identity, Impenetrability, and the Desideratum of Particle Number Conservation in TBM*

In addition to the mere permutations, it is argued that we also have to remove the diagonals where particles coincide (ibid., 231), i.e., "all points corresponding to two or more particles occupying the same spatial position at the same instant", resulting in the further reduced configuration space $\mathbb{R}^{Nd} / S_N - \Delta$. As noted by Brown et al. (ibid.), "[this] removal […] is sometimes justified by considering the particles to be 'impenetrable', but of course impenetrability is not a direct consequence of indistinguishability." Rather, as French and Krause (2006, 40) argue, referring to Newton, "impenetrability is a fundamental assumption of classical physics and we find it explicitly stated in a number of basic texts. Thus Newton characterized the elements of his 'proto'-theory of classical mechanics in terms of (1) impenetrability, (2) mobility and (3) the ability to excite the senses." Nevertheless, impenetrability is a metaphysical assumption that might resonate with our intuitions, but needs further justification once we want to make use of it in the construction of our theories about nature. Consequently, French and Krause (ibid., 41) turn the argument around, stating that "for point particles in three dimensions, the regions of coincidence in phase space have zero measure, so impenetrability is a reasonable assumption." Brown et al. (1999, 231) are calling this line of argument "standard wisdom" when it comes to the construction of reduced configuration spaces, referring to Wu (1984, 2104), who argues that "because we are not always guaranteed that there is a finite probability for two particles to coincide with each other, the so-called diagonal points in $\mathbb{R}^{dN}$ [...] have to be excluded too."

However, it is questionable whether this argument from the non-finite probability of particle coincidences, and likewise its measure-theoretic analogue brought forward by French and Krause, can convincingly justify the assumption that particles are impenetrable: belonging to a region of zero measure and hence being assigned a non-finite probability is not to be equated with being impossible. The simple case of a system consisting of only one point-like particle should make that clear. Wherever this particle is located, the region it occupies is of measure zero, unless we assume a coarse-grained cell structure with cells of finite measure for the definition of regions in configuration space, assigning a finite region to the particle's location. Consequently, without imposing such coarse graining, the particle has probability zero of being where it is. Yet, it is there. Clearly, being assigned probability zero or having measure zero are not the same as being impossible. The scope of this argument is not limited to the one-particle case. It holds just the same for two or more particles coinciding, and in general, whenever we are dealing with sets of configurations that have zero measure/are assigned non-finite probability. These sets don't even have to be small, despite their zero measure. The diagonal in question, for example, has the same cardinality as the whole of configuration space.

But then, if these configurations are neither impossible a priori nor of negligible number, why exclude them from configuration space? In a footnote, Wu (1984, fn. 12) points to the fact that, if we didn't exclude the diagonal, we would only be able to obtain Bose-Einstein statistics. This type of argument is different from the ones we have encountered so far. The question of which statistics is adequate for the phenomena to be described is one that can be tackled empirically. If we have to reduce configuration space by the diagonal in order to arrive at a model that captures the phenomena, then we should do so. On these rather solid grounds, then, we may start building a suitable metaphysics for that model, arguing, for example, that the fact that the model, by not allowing for the coincidence of particles, makes correct statistical predictions, should be taken as evidence that the particles in question are impenetrable.

Yet another argument for the subtraction of the diagonal, probably the most important one, goes as follows. Let's say two particles inhabit the same point in space-time. Assume that these particles have no discerning features, such that they are rendered identical by the above PII. Then, because of them coinciding, they are rendered identical even with respect to their spatiotemporal location, resulting in numerical identity, unless we assume haecceity.[31] So, instead of two particles, we'd

[31] Numerical discernibility of coinciding, identical particles is lost, because they are not even weakly discernible, which would be sufficient for them being numerically discernible. Saunders (2003, 293f.), referring to Black's thought experiment of two iron spheres (Black 1952), provides the irreflexive relation 'x being one mile apart from y' as an example for a weakly discerning feature: "An identity condition may fail even when objects have exactly the same properties and

end up with only one, either featureless, or, given the original particles had features at all, e.g., mass, with one particle that has double the mass. In any case, our 2-particle system would now consist of only one particle, a situation described by a completely different model ($\mathcal{Q}$, WF, SE) in the first place. This inconsistency could only be resolved by assuming primitive thisness/haecceity, which would allow for the preservation of the number of particles despite their coincidence. But if we are assuming haecceity in the case of coincident particles in order to keep N constant, why don't we assume haecceity for particles in general? In fact, I think we would have to, not least because we don't know which particles actually are going to coincide in a future configuration, or did coincide in an earlier one. In standard BM, we do know these things, since there, particles are assumed to travel on trajectories given by the first order GE, which tells us that particles coinciding at one time do so at all times—and conversely, that particles not coinciding at one time never do. Therefore, in standard BM, once we know which particles are coinciding and which aren't, our knowledge about that fact is stable. Not so in TBM, where, in principle, there's nothing to prevent particles from coinciding at one time and not coinciding at others. So, in order to keep N constant, we would need to assume haecceity for all particles. Otherwise it would be possible that two or more non-haecceitistic particles coincide, thereby reducing N. Hence, we really should assume haecceity for particles in general, if we make that assumption at all. But there's a catch: once we assume haecceity for all particles, the reduced configuration space is no longer appropriate to model the N-particle system in question. We'd have to use the unreduced configuration space, because particle permutations amount to different facts about the world, qua the haecceity of particles. This renders the initial reason for the assumption of haecceity absurd. We considered this option in order to be able to keep the diagonal while reducing configuration space. But if we make the assumption, we block the possibility of reduction in general, with or without the subtraction of the diagonal. So, once we assume haecceity *in order to* be able to reduce configuration space in a certain way, we can't reduce at all. Therefore, assuming haecceity for that rather formal purpose is a non-starter. Rather, particle haecceity, being an independent, metaphysical assumption, requires strong justification of its own. Whether such justification can be given I shall discuss now.

4.7.2 Particle Haecceity vs. Haecceitism

Following Cowling (2016), haecceitism is the position "according to which the world could differ non-qualitatively without differing qualitatively." For an example related to our discussion, consider, in analogy to Black's iron spheres, a universe

exactly the same relations to all other objects *and* exactly the same relations to each other; [...] Call objects not absolutely or relatively discernible, that satisfy an irreflexive relation, *weakly discernible*; [...] For an example of weakly discernible objects, consider Black's two iron spheres, one mile apart, in an otherwise empty space. [...] The irreflexive relation A is '... one mile apart from ...'. It is *because* this relationship holds that we may say that there are two [...]".

consisting of two intrinsically identical particles at some spatial distance. This, in the example, is the actual world, and as such, a possible one. Now imagine another possible world where these two particles have swapped places. Haecceitism is the claim that these two worlds exist in the space of possible worlds, and that these possible worlds are different, despite being qualitatively the same.

Based on this characterization, I'd like to distinguish *haecceitism* from *particle haecceity*, the claim that particles *actually* differ in spite of sharing all of their properties, i.e., that the respective particles do have some non-qualitative individuating features to them. Haecceitism with respect to particles, as I am using the notion, merely makes the weaker claim that it is *possible* that they do. At this point, it should be made clear again that we are interested in whether or not we have good reasons to assume particle haecceity, not so much whether particle haecceity, in principle, is possible, i.e., whether or not we have good reasons to assume haecceitism. This complicates the issue to some extent. Chisholm's Paradox, for example, as Cowling (2016) notes, is one of the most prominent arguments brought forward in favour of haecceitism. Unfortunately, though, it is of not much use for our purposes, for it seems to presuppose the individuality of entities in question and their trans-possible-world identity (Chisholm, 1967, 3): "The [entity] of this world, we are assuming, is identical with the [entity] of that one" and "moving from one possible world to another, but keeping our fingers, so to speak, on the same two entities [...]" are just two quotes that point towards this presupposition. In more formal presentations of Chisholm's Paradox, this issue shows, appropriately, in the formalism. For example, the two-valued predicate M in the presentation given in Forbes (1984) assigns some set of properties or constituents h_i to some entity α. While h_i changes, thereby representing change of the set of properties or constituents assigned to the entity, α remains the same over the entire operation, the latter being the switching between possible worlds. It is, by assumption, always the same, individuated entity, through all possible worlds. But for our purposes, we must not allow ourselves to make such assumptions, for what is assumed here is exactly the subject matter of the open question we are discussing in this paragraph. Simply assuming individuality and/or trans-world identity would be begging the question.

Other arguments in favour of haecceitism, according to Cowling (2016, 4.1), are mostly of the conceivability type, characterized as follows: "Conceivability arguments for haecceitism have two steps. The first step requires our success in conceiving or imagining certain states of affairs. The second step requires an inference from the relevant conceiving or imagining to the possibility of the states of affairs in question." We are not going to discuss these types of arguments here in detail, for they do not promise to be exceptionally relevant for the special case we are interested in. Generally speaking, when it comes to the question of the haecceity of particles, conceivability type arguments roughly will go as follows: (P1) It is conceivable that two (spatio-temporally identical) particle configurations differ, despite the particles in question being qualitatively identical. (P2) If something is conceivable, it is possible. (C) It is possible that two (spatio-temporally identical) particle configurations differ, despite the particles in question being qualitatively identical.

Now, regardless of the actual truth-value of P1 and P2, assuming they are true, C is true as well. Nevertheless, the argument has no bearing on our issue at hand. It might well be that configurations differing only with respect to the primitive identity of their particles are conceivable, and it also might well be that, since they are conceivable, that they are possible. But the question we have been trying to answer is different. It is: do particles actually *have* primitive identity? Is their fundamental nature *in fact* haecceitistic? We wanted to be presented with good reasons to assume particle haecceity. What we got instead was the assertion that, given the assumption of some premises, haecceity is possible, i.e., an argument in favour of *haecceitism*, understood as the *possibility of haecceity*—a conviction we shared from the beginning.

We will not ultimately resolve the issues surrounding particle haecceity and general haeccetism here. If the reader feels justified to assume particle haecceity for Bohmian particles, for them, the issue of particle number conservation of coinciding, identical particles in TBM is off the table. Yet, I am unable to find convincing arguments in favour of the haecceity of particles. Accordingly, I have no justification to make use of haecceity in order to solve the problem. So, particle number conservation remains an issue. Maybe we can find a solution by investigating so-called space-time individuality (STI) of particles.

4.7.3 Individuation of Particles Qua their Space-Time Trajectories

Earlier, I have mentioned the possibility of particle individuation qua space-time trajectories. Again, I shall make use of an illustrative example. Imagine a universe with all identical particles. At t_1, we find the particles in configuration Q_1. At a later time t_2, we find them in Q_2. Now, let's assume that Q_1 and Q_2 are identical, in the sense that all points in space that have been occupied by a particle at t_1 are equally occupied at t_2. Since particles are identical and thus not discernible through their intrinsic properties, the configurations, too, are indiscernible, despite the particles having changed places. Yet, so the argument goes, there is a way to tell these particles, and a fortiori the configurations, apart: In general, on their way from Q_1 to Q_2, particles were travelling on different trajectories. Once we know these trajectories, we can regard them as a feature of the particles that renders them discernible. French and Krause (2006, 1.7) reconstruct the conditions under which such a trans-temporal re-identification of particles via their trajectories can be successful. As the first amongst these conditions, they mention spatio-temporal continuity of the trajectory. And some pages later, in their discussion of classical statistical mechanics (ibid., 2.1), they describe the analogue issue that shows there:

> Given that the molecules are all indistinguishable in the sense of possessing the same intrinsic or state-independent properties, some further principle of individuality is required in order to even talk about distinct trajectories in the first place. Boltzmann, more philosophi-

> cally minded than many of his contemporaries, was completely aware of this issue. In his treatise on the principles of mechanics, the very first axiom of mechanics states that indistinguishable particles which cannot occupy the same point of space at the same time can be individuated by the initial conditions of their trajectories and the continuity of their motion. This, Boltzmann insisted, '... enables us to recognize the same material point at different times'.

This position is what French and Krause call space-time individuality (STI), and, as they note already in the introduction (ibid., 12), "it immediately invites speculation as to the nature of the individuating spatio-temporal background. On a relational view of space and time, the spatio-temporal locations of physical individuals involve relations with other individuals and a possible circularity develops. This may be broken by invoking a privileged set of continuants by reference to which all other things are individuated. But the nature of this privilege must be spelled out and the danger is that STI collapses into some other account of individuality [...]. Alternatively, the circularity could be broken by adopting a substantivalist view of space and time according to which spatio-temporal properties and relations of physical things are reduced to the properties of space-time points. Of course, this in turn requires some explication of what it is that confers individuality upon the points of space-time themselves."

We don't have to delve into these further issues here, for it already should be clear why STI, in our case, can't be utilized for the individuation of particles: there simply are no continuous particle trajectories in TBM, so the necessary condition of STI isn't met. Particle individuation via space-time trajectories, thus, isn't a possible solution to our problem.

So, what are we left with? First, as we have just seen, we can't make use of STI to individuate otherwise identical particles within the TBM framework. Second, while not being strictly impossible, particle haecceity isn't a promising candidate either, for there are no convincing arguments in favour of this assumption.

Yet, there is the formal requirement within the framework of TBM that particle number be conserved, i.e., that N is constant over time, which, without the assumption of haecceity, would be violated once particles do coincide. This requirement is regarded as a postulate of the framework, and cannot simply be dropped, for reasons discussed above. (Recall that two systems with different particle numbers would be described by different models.) The reasonable way out seems to be to assume that particles in TBM indeed cannot coincide, i.e., are impenetrable. Note that this does not contradict the remarks about the relationship between indistinguishability and impenetrability by Brown et al. (1999). Indeed, "impenetrability is not a direct consequence of indistinguishability." (231) Instead, it is a consequence of (1) the formal desideratum of particle number conservation, on the one hand, and of (2) indistinguishability *plus* (3) the rejection of particle haecceity, on the other. Only if we want all three, we seem to be forced to assume impenetrability: once particles have haecceity or are distinguishable, impenetrability is not necessary for particle number conservation. But this is true only if *all* particles are distinguishable, given, as we have argued, haecceity is not a reasonable option: as soon as some particles are indistinguishable, conservation of particle number is prone to failure due to possible particle coincidence. Consequentially, impenetrability must be assumed as soon as two or more particles of a system are indistinguishable, even if we can distinguish particle *types* based on their intrinsic features. To avoid particle coincidences leading to variations in N, we can reduce configuration space not only by the main diagonal, i.e., the configurations where all particles coincide, but also by the secondary diagonals, i.e., the diagonals of the configuration sub-spaces describing the configurations involving two or more, but not all, indiscernible particles. This seems appropriate for modelling a universe that is made up of different, distinguishable particle types, yet allows for indistinguishable particles within those types.

Conclusively, we can say that reduction of configuration space by the diagonals indeed is a formal reflection of the metaphysical assumption of impenetrability. But, as we have seen, at least in TBM, the reasoning behind this assumption is not as trivial as claiming it to be "standard wisdom" or simply referring to our intuitions. Rather, it is a requirement necessary for other desiderata to hold—particle number conservation and no particle haecceity—and only so in the case of indistinguishable particles.

Now that we have gained some clarity about the fundamental space we are operating in, and about the nature of its inhabitants, we finally can approach the ultimate definition of worlds in TBM and see how world histories are construed.

4.8 What Is a World? Redux

A particle configuration $Q_i \in \mathbb{R}^{3N}$ is a point given by the positions of the N particles of a given universe. A world $\mathcal{W}_{i(MO)} \in \mathbb{R}^{3N}$ is an area within the configuration space around a configuration Q_i, containing all possible configurations that, for a given MO within Q_i, satisfy SS with Q_i, i.e., that are sufficiently similar to Q_i, i.e., that are macroscopically indistinguishable from Q_i in the sense that change between them doesn't affect the MO macroscopically, i.e., doesn't make a macroscopic difference to it.[32] So let's alter our world membership condition to reflect the MO's relevance:

$$\forall \mathcal{Q}_{i(MO)} \exists \mathcal{W}_{i(MO)} \left(\forall Q_j \left(Q_j \in \mathcal{W}_{i(MO)} \leftrightarrow Q_j SS_{(MO)} \mathcal{Q}_{i(MO)} \right) \right)$$

(*world membership*$_{\text{TBM}}$)

Using this condition, we can define the world $\mathcal{W}_{i(MO)}$ around $\mathcal{Q}_{i(MO)}$, containing the MO in question, as the set of all (possible) configurations that, for the MO, are

[32] Of course, it is difficult to even make sense of the notion of macroscopic objects in a toy universe comprising only two particles in one spatial dimension. This toy universe only serves illustrational purposes, to render the underlying principle of the construal of worlds from configurations clear. They are equally applicable to universes with many more particles with more spatial degrees of freedom.

sufficiently similar to $\mathcal{Q}_{i(MO)}$ in that their actualization doesn't amount to macroscopic change:

$$\mathcal{W}_{i(MO)} = \left\{ Q_j | Q_j SS_{(MO)} \mathcal{Q}_{i(MO)} \right\}$$

This world trivially contains $Q_{i(MO)}$, since the latter is trivially sufficiently similar to itself.

For defining the world around an *actualized* configuration at a given time, we naturally take for $Q_{i(MO)}$ the configuration actualized at that time.

Note that not necessarily every possible configuration in configuration space has a world around it thus defined. There are only worlds where there are MOs, because it depends on the MO what counts as sufficiently similar (or not) to the actualized configuration. On the other hand, if there is an MO, then there is a world, and if there are several MOs, then there are several worlds. Furthermore, if a configuration contains several MOs, macroscopically being affected by different degrees of microscopic change, differently sized worlds emerge from that one configuration.

4.9 From Discrete Configurations to Continuous World Histories via Overlapping Worlds

We shall now tackle the distinction between overlap simpliciter and overlap proper, and show why only the latter is sufficient for world histories to emerge. Let's start by sharpening our intuitions with respect to overlapping worlds with the help of some illustrations. The crucial feature of worlds as defined above is that they extend through configuration space. Consequently, *worlds can overlap without being identical*.

Figure 4.3 shows worlds $\mathcal{W}_i$ and $\mathcal{W}_k$ overlapping in configuration space, with $Q_i \notin \mathcal{W}_k$ and $Q_k \notin \mathcal{W}_i$. This is what we call *overlap simpliciter*. From the

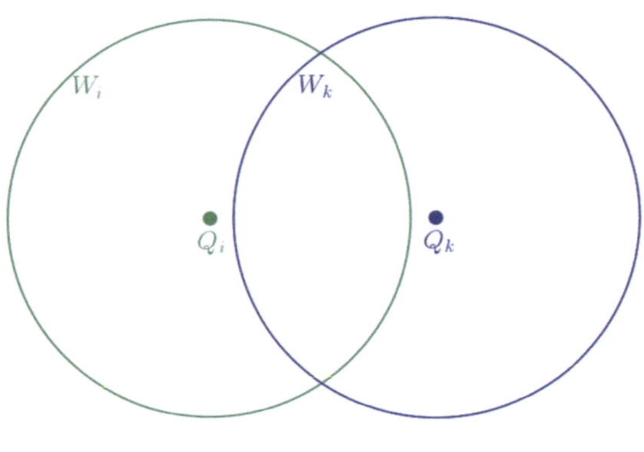

Fig. 4.3 Two worlds overlapping

perspective of the MO in Q_i, the actualization of Q_k would amount to the instantiation of a state differing macroscopically from the one instantiated by Q_i, and vice versa. In this scenario, no (FAPP-) continuous world history would emerge for the MO. $\mathcal{W}_i$ and $\mathcal{W}_k$, despite their overlap, in this scenario are two quasi-parallel[33] worlds.

Only when there is another configuration Q_j actualized within the overlap of $\mathcal{W}_i$ and $\mathcal{W}_k$ can a world history emerge, and FAPP-continuous macroscopic change happen:

Figure 4.4 shows the same scenario as Fig. 4.3, but now with another configuration Q_j that is actualized in the overlap of $\mathcal{W}_i$ and $\mathcal{W}_k$, and that includes both Q_i and Q_k in its world $\mathcal{W}_j$. Now, a FAPP-continuous world history is possible: Let's say we start from Q_i. Q_j lies within $\mathcal{W}_i$, so if it is actualized, from the perspective of the MO in Q_i, no macroscopic change happens, despite the fact that now, a different configuration Q_j is actualized and hence, a different world $\mathcal{W}_j$ instantiated. Q_i and Q_j satisfy SS, so the MO, from its own macro perspective, persists through the jump between $\mathcal{W}_i$ and $\mathcal{W}_j$. The same holds for the jump between $\mathcal{W}_j$ and $\mathcal{W}_k$: now residing in Q_j, the MO is not affected by any change when Q_k is actualized. As before, strictly speaking from the fundamental level, the configuration changes; and in accordance with that, different worlds are instantiated. But from the MO's macroscopic perspective, everything remained the same, since Q_j and Q_k satisfy SS.

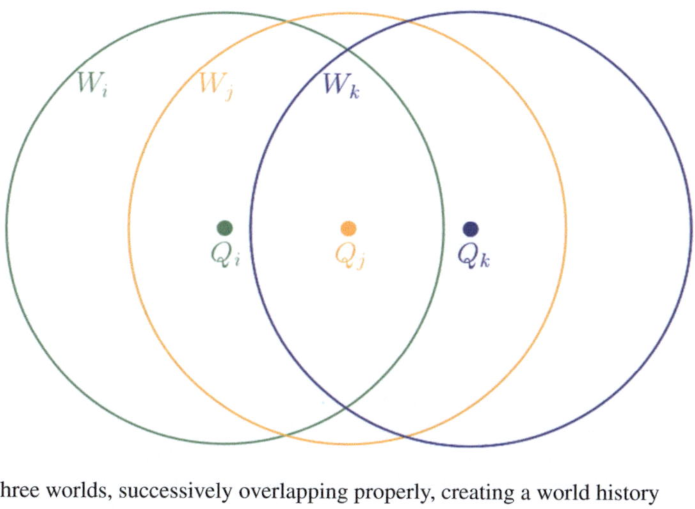

Fig. 4.4 Three worlds, successively overlapping properly, creating a world history

[33] "Quasi-parallel" because in TBM, configurations and worlds are actualized/instantiated sequentially, so worlds, and likewise world histories, do not exist simultaneously, as they do in Everett-style accounts.

This is essential. During the configurational jump from Q_i to Q_j, the MO underwent no macroscopic change. Likewise, during the jump from Q_j to Q_k. It macroscopically persists through all jumps, although it doesn't persist on the fundamental level: it quasi- or FAPP-persists. Yet, judging from the perspective of the MO in Q_i, Q_k does not lie within its world $\mathcal{W}_i$, so overall, with the actualized configuration going from Q_i to Q_k via Q_j, the MO did undergo macroscopic change. Instead of the instantiation of two quasi-parallel worlds $\mathcal{W}_i$ and $\mathcal{W}_k$, the actualization of Q_j qua instantiating its world $\mathcal{W}_j$ links up $\mathcal{W}_i$ and $\mathcal{W}_k$ in a world history $\{\mathcal{W}_i, \mathcal{W}_j, \mathcal{W}_k\}$, with Q_i, Q_j, Q_k as their respective central configurations. It is only through such chains of actualizations of sufficiently similar configurations, linking up configurations that are not sufficiently similar, that MOs undergo FAPP-continuous, macroscopic change.[34]

Let's make all this a bit more precise. Recall our definition of a world:

$$\mathcal{W}_{i(MO)} = \{Q_j | Q_j SS_{(MO)} Q_{i(MO)}\}$$

We now need to know when worlds containing MOs overlap. They do when their intersection is non-empty: $\mathcal{O}_{\mathcal{W}_{i(MO),j(MO)}} = \mathcal{W}_{i(MO)} \cap \mathcal{W}_{j(MO)} \neq \varnothing$. Recall our characterization of overlapping worlds in BSM. There, we said that a microstate x_k is in an overlap $\mathcal{O}_{\mathcal{W}_{i,j}}$ of worlds $\mathcal{W}_i$ and $\mathcal{W}_j$ iff it is in both of these worlds:

$$\forall x_k \forall \mathcal{W}_{i,j} \left(x_k \in \mathcal{O}_{\mathcal{W}_{i,j}} \leftrightarrow x_k \in \mathcal{W}_i \wedge x_k \in \mathcal{W}_j \right)$$

We can easily adapt this to TBM's configurations:

$$\forall Q_{k(MO)} \forall \mathcal{W}_{i(MO),j(MO)} \left(Q_{k(MO)} \in \mathcal{O}_{\mathcal{W}_{i(MO),j(MO)}} \leftrightarrow Q_{k(MO)} \in \mathcal{W}_{i(MO)} \wedge Q_{k(MO)} \in \mathcal{W}_{j(MO)} \right)$$

However, this only tells us about overlap simpliciter, i.e., about the necessary and sufficient condition for a micro state to be in an overlap region. For our treatment of BSM, since we were dealing with a continuous trajectory of the micro state through

[34] Isn't this surprisingly reminiscent of Wittgenstein's account of "family resemblance" in his Philosophische Untersuchungen? (Wittgenstein 1953, §67, my translation): "And we are expanding our notion of number, just as we twist fibre on fibre when spinning a thread. And the strength of the thread does not lie in one fibre running through its whole length, but in many fibres overlapping."

Of course, instead of „expanding our notion of number", we are expanding world histories, and the family resemblance we make use of is the relation SS between configurations: We are twisting "fibre on fibre" by having properly overlapping worlds instantiated, and the strength of the resulting threat, i.e., the world history, does not "lie in one fibre running through its whole length"—one continuous trajectory of the configuration—"but in many fibres [(i.e., worlds)] overlapping".

phase space, this was suitable. In such a scenario, the next state instantiated lies arbitrarily close to its predecessor, which guarantees that it is in the latter's world and that thus, the worlds instantiated overlap properly. Yet, for the discontinuous micro dynamics of TBM, we don't have such a guarantee. Hence, overlap simpliciter is not enough. Worlds must overlap properly, i.e., in such a way that their respective central configurations lie within the world around the other central configuration: $Q_i \in \mathcal{W}_j \wedge Q_j \in \mathcal{W}_i$. That is to say, worlds do overlap properly when their respective central configurations lie within their overlap, i.e., within the intersection $\mathcal{O}_{\mathcal{W}_{i(MO),j(MO)}} = \mathcal{W}_{i(MO)} \cap \mathcal{W}_{j(MO)}$.

Two worlds $\mathcal{W}_{i(MO)}$ and $\mathcal{W}_{j(MO)}$ overlap properly iff the respective central configurations $Q_{i(MO)}$ and $Q_{j(MO)}$ lie within the overlap of these worlds. To express this, we write $\mathfrak{D}_{\mathcal{W}_{i(MO),j(MO)}}$:

$$\forall Q_{i(MO),j(MO)} \forall \mathcal{W}_{i(MO),j(MO)} \left(\mathfrak{D}_{\mathcal{W}_{i(MO),j(MO)}} \leftrightarrow Q_{i(MO),j(MO)} \in \mathcal{O}_{\mathcal{W}_{i(MO),j(MO)}} \right)$$

Having specified proper overlap, we can now construe world histories. Two worlds, despite overlapping properly, don't make a world history, for the simple reason that jumping from one central configuration to the other can, by definition, never amount to macroscopic change for the MO. We only arrive at a world history, e.g., $\mathfrak{H}_{\mathcal{W}_{i,j,k}}$, once we have proper overlap between worlds $\mathcal{W}_i, \mathcal{W}_j$ and proper overlap between worlds $\mathcal{W}_j, \mathcal{W}_k$ but not between $\mathcal{W}_i, \mathcal{W}_k$. There can be more worlds overlapping properly in between, but they don't contribute to the MO undergoing macroscopic change. Also, $\mathcal{W}_i, \mathcal{W}_k$ may overlap simpliciter, but not properly. So, we want to define a world history as a set of worlds such that all neighbouring worlds properly overlap while the first and last world do not:[35]

$$\mathfrak{H}_{\mathcal{W}_{i,\dots,n(MO)}} = \left\{ \mathcal{W}_{i(MO)}, \mathcal{W}_{j(MO)}, \dots, \mathcal{W}_{m(MO)}, \mathcal{W}_{n(MO)} \middle| \mathfrak{D}_{\mathcal{W}_{i(MO),j(MO)}} \wedge \mathfrak{D}_{\mathcal{W}_{j(MO),\dots}} \wedge \dots \wedge \mathfrak{D}_{\mathcal{W}_{\dots,m(MO)}} \wedge \mathfrak{D}_{\mathcal{W}_{m(MO),n(MO)}} \wedge \neg \mathfrak{D}_{\mathcal{W}_{i(MO),n(MO)}} \right\}$$

Read: the MO's world history $\mathfrak{H}_{\mathcal{W}_{i,\dots,n(MO)}}$ is the set of worlds $\mathcal{W}_{i(MO)}$ through $\mathcal{W}_{n(MO)}$ such that consecutive pairs of worlds overlap properly, but not the first and the last world.

Finally, we have arrived at an account of macroscopically FAPP-continuous change emerging from fundamentally discontinuous actualizations of configurations via the construction of world histories from properly overlapping worlds around individual configurations containing MOs. Note that this FAPP-continuous

[35] Order of the set of worlds in a world history is implicit via the imposed overlap conditions, such that we don't need to construe the set of worlds in a world history as an ordered set explicitly.

change throughout a world history in no way necessitates continuous change between configurations. We arrived here starting from the framework of TBM without applying any fundamental changes. In particular, the discontinuous, random actualization of configurations in configuration space, which arguably is the central and discerning feature of TBM, has not been altered.

A consequence of the emergent nature of macroscopically continuous change, TBM comes with the feature that world histories can be interrupted "externally" without being interrupted "internally". That is to say: in principle, between two actualizations that belong to one world history as just described, there can lie arbitrarily many actualizations that do not belong to that world history. It could be the case, e.g., that in the time between Q_{j1} and Q_{k1} from $\mathfrak{H}_{W_{i1,\dots,n1(MO)}}$ are actualized, Q_{j2} through Q_{l2} from a different world history $\mathfrak{H}_{W_{i2,\dots,n2(MO)}}$ are actualized. In fact, arbitrarily many configurations, belonging to arbitrarily many different world histories, can be actualized in between the actualization of two consecutive configurations of a given world history. The MO in the latter wouldn't be affected. Such, I call an external interruption. The world history can be taken up afterwards without noticing, for all individual configurations are static and actualized instantaneously anyway. Internally, world histories are insensitive to external interruption, as long as the conditions of their worlds overlapping properly are met.

How, then, is this proper overlap between the worlds of individual configurations guaranteed, such that a FAPP-continuous world history can emerge? Here comes into play our treatment of the actualization rate of configurations within a world in a finite time interval (cf. 2.7): with configurations being actualized at an arbitrarily high rate, the expectation of configurations belonging to a certain world being instantiated during the same time interval becomes arbitrarily high as well. Apply this to the issue of emergent world histories being FAPP-continuous: given arbitrarily many actualizations within a finite time interval, a configuration being sufficiently similar to a given configuration is practically guaranteed to be actualized within an arbitrarily small amount of time. Once it is actualized, another configuration, being sufficiently similar to the second one, is actualized after an arbitrarily small amount of time, and so on. Always lying within the world of the previous configuration, the consecutive worlds thus instantiated always overlap properly, guaranteeing that world histories continue. It hence is unreasonable to fear that your world history will end just because no configuration within the world of the presently actualized configuration will be actualized "in time" (as in "punctually").

4.9.1 Relative Frequency of World Histories?

Despite, the relative frequencies with which macroscopically different world histories are instantiated may very well differ in value. Given an arbitrarily high yet finite overall actualization rate, we can, in principle, count the number of actualizations within world histories resembling each other with respect to a certain macroscopic feature during a finite time interval and build a quotient with the overall number of actualizations during that time interval, thus providing a way to "measure" the probabilities of such world histories to be instantiated in terms of frequencies. This is not to say, though, that TBM's probabilities are defined in terms of relative frequencies. Likewise, this must not be confused with counting branches, or counting world histories. World histories in TBM, like world branches in Everett, cannot be counted, strictly speaking. They can be FAPP-counted, but more on that below (4.9.5). Instead, these relative frequencies are merely *reflecting* the (dispositional) weight of Everettian world branches (cf. Wallace, 2012, Chap. 4), providing, so to speak, a specific frequentist way of assessing this weight that is not open to the pure Everettian, who thinks of world branches as being instantiated simultaneously, with their overall combined weight being unity. In short, we are not committing to frequentism with respect to probabilities in the usual sense, but can use it as a tool to assess dispositional weights of world branches/propensities of world histories to be actualized.

To flesh this out a little, consider an experiment with two possible, mutually exclusive, macroscopic outcomes e and $\overline{e}$, with probability $\mathrm{P}(e)=\frac{2}{3}$ and probability $\mathrm{P}(\overline{e})=1-\mathrm{P}(e)=\frac{1}{3}$. That is to say, in Everettian terminology, the combined weight of all branches in which e occurs is $\frac{2}{3}$, and the combined weight of all branches in which $\overline{e}$ occurs is $\frac{1}{3}$. For TBM, this translates as follows. If we assume some sufficiently high but finite actualization rate in a given time interval, then the totality of world histories in which e occurs is instantiated at a relative frequency of about $\frac{2}{3}$ of all actualizations in the time interval considered. Likewise, the totality of world histories in which $\overline{e}$ occurs is instantiated in about $\frac{1}{3}$ of all actualizations in that time interval. This is merely to convey a sense of what it means for world histories to lie in the support of world branches with certain weights, spelled out in terms of relative frequencies of actualizations: world histories in the support of world branches endowed with greater weights are instantiated "more often" than world histories in the support of world branches endowed with lesser weights.

Do these differences in relative frequencies, then, mean that world histories in the support of world branches endowed with very low weights practically cannot emerge? Such fear amounts to a confusion of relative frequency and actualization rate. While the former is approximately stable over all sufficiently high actualization rates—i.e., relative frequencies approximately stay the same, independently of the actualization rate, as long as the latter is sufficiently high—the actualization rate itself can take on arbitrarily high values, such that any world history is actualized arbitrarily often, no matter how low the weight of the respective world branch. And of course, it should be stated explicitly what is already implicit in the assumptions we made about the actualization rate: all this talk of relative frequencies only makes sense given the overall actualization rate is finite. If we assume an infinite overall actualization rate, as we generally do on TBM, we run into the problems familiar from accounts of probabilities in terms of "frequencies" in infinite series of events. For example, assume an infinite overall actualization rate, such that in any finite time interval, infinitely many configurations are actualized. Then, any world history is instantiated infinitely many times in the same time interval. How to calculate, from these infinites, a relative frequency? It might be helpful to imagine the following: If we were able to look at the universe as described by TBM from the outside, but did so only a finite, but sufficiently large number of instants in a given time interval, we'd expect to approximately recover the relative frequencies as above. If, on the other hand, we don't restrict ourselves to looking only a finite number of instants in that time interval, we don't recover relative frequencies.

So much for now about relative actualization frequencies and their role in TBM. We must now come back to world histories, and the ways in which they populate configuration space.

4.9.2 *Quasi-Parallel World Histories*

Above, we have described worlds that do not overlap properly as quasi-parallel worlds. Quasi-parallel worlds can be linked up by world histories. In fact, as argued above, if world histories are to account for macroscopic change, it is crucial that two of their worlds do not overlap properly. Otherwise, no macroscopic change would emerge, since all actualisations within that history would lie within the worlds of all other actualisations. It is for this reason that we included the condition—that two worlds within one world history must not properly overlap—into the definition of a world history. The linking-up of two worlds not properly overlapping is the whole point of world histories, to put it another way.

Just as worlds are said to be quasi-parallel if they do not overlap properly, we can define a condition under which world histories are quasi-parallel. The idea of defining quasi-parallel world histories as those world histories which do not overlap properly suggests itself. But what does it mean for world histories to not overlap properly? Is it necessary that no world from one world history, $\mathfrak{H}_{\mathcal{W}_{i1,\ldots,n1}}$, say, properly overlaps with a world from another world history $\mathfrak{H}_{\mathcal{W}_{i2,\ldots,n2}}$? Or will it be sufficient if at least one world each does not overlap properly with a world from the other world history, respectively?

To get a grip on this, let's start with the simplest case, in which no world contained in one world history properly overlaps with a world from another world history. First, we need two world histories:[36]

[36] For sake of readability, from now on we will simply write $\mathfrak{H}_1$ instead of $\mathfrak{H}_{\mathcal{W}_{i1,\ldots,n1(MO)}}$ etc., and leave out explicit reference to the MO in indices, always keeping in mind that world histories consist of worlds, and worlds only arise from configurations containing MOs.

$$\mathfrak{H}_1 = \{\mathcal{W}_{i1}, \mathcal{W}_{j1}, \ldots, \mathcal{W}_{m1}, \mathcal{W}_{n1} \mid \mathfrak{O}_{\mathcal{W}_{i1,j1}} \wedge \mathfrak{O}_{\mathcal{W}_{j1,\ldots}} \wedge \ldots \wedge \mathfrak{O}_{\mathcal{W}_{\ldots,m1}} \wedge \mathfrak{O}_{\mathcal{W}_{m1,n1}} \wedge \neg \mathfrak{O}_{\mathcal{W}_{i1,n1}}\}$$

and

$$\mathfrak{H}_2 = \{\mathcal{W}_{i2}, \mathcal{W}_{j2}, \ldots, \mathcal{W}_{m2}, \mathcal{W}_{n2} \mid \mathfrak{O}_{\mathcal{W}_{i2,j2}} \wedge \mathfrak{O}_{\mathcal{W}_{j2,\ldots}} \wedge \ldots \wedge \mathfrak{O}_{\mathcal{W}_{\ldots,m2}} \wedge \mathfrak{O}_{\mathcal{W}_{m2,n2}} \wedge \neg \mathfrak{O}_{\mathcal{W}_{i2,n2}}\}$$

Above, we have specified proper overlap for worlds within a world history. We can likewise specify proper overlap for worlds from different world histories $\mathfrak{H}_1$, $\mathfrak{H}_2$:

$$\forall \mathcal{Q}_{i1,i2} \forall \mathcal{W}_{i1,i2} \left(\mathfrak{O}_{\mathcal{W}_{i1,i2}} \leftrightarrow \mathcal{Q}_{i1,i2} \in \mathcal{O}_{\mathcal{W}_{i1,i2}} \right)$$

We can now differentiate between four basic ways in which world histories relate to each other:

We say that world histories are *entirely quasi-parallel* iff none of their respective worlds overlap properly with any world from the other world history. For this, we write $\mathfrak{H}_1 ::: \mathfrak{H}_2$:

$$\mathfrak{H}_1 ::: \mathfrak{H}_2 \leftrightarrow \neg \exists \mathcal{W}_{i1,i2} \left(\mathfrak{O}_{\mathcal{W}_{i1,i2}} \right)$$

(*entirely quasi-parallel*)

Accordingly, we say that world histories are *entirely overlapping* ($\mathfrak{H}_1 \ldots \mathfrak{H}_2$) iff each world of the first world history overlaps properly with some world from the other world history *and* vice versa:

$$\mathfrak{H}_1 \ldots \mathfrak{H}_2 \leftrightarrow \forall \mathcal{W}_{i1} \left(\exists \mathcal{W}_{i2} \left(\mathfrak{O}_{\mathcal{W}_{i1,i2}} \right) \right) \wedge \forall \mathcal{W}_{i2} \left(\exists \mathcal{W}_{i1} \left(\mathfrak{O}_{\mathcal{W}_{i1,i2}} \right) \right)$$

(*entirely overlapping*)

Word histories are *partially quasi-parallel* ($\mathfrak{H}_1 .:. \mathfrak{H}_2$) iff not every world from the first world history overlaps properly with a world from the other world history *or* vice versa:

$$\mathfrak{H}_1 .:. \mathfrak{H}_2 \leftrightarrow \neg \forall \mathcal{W}_{i1} \left(\exists \mathcal{W}_{i2} \left(\mathfrak{O}_{\mathcal{W}_{i1,i2}} \right) \right) \vee \neg \forall \mathcal{W}_{i2} \left(\exists \mathcal{W}_{i1} \left(\mathfrak{O}_{\mathcal{W}_{i1,i2}} \right) \right)$$

(*partially quasi-parallel*)

Partially overlapping ($\mathfrak{H}_1 :.: \mathfrak{H}_2$) world histories are such that there are worlds from these different world histories that overlap properly:

$$\mathfrak{H}_1 :.: \mathfrak{H}_2 \leftrightarrow \exists \mathcal{W}_{i1,i2} \left(\mathfrak{O}_{\mathcal{W}_{i1,i2}} \right)$$

(*partially overlapping*)

These basic relations between world histories can be arranged in a diagram I call the square of world histories, in analogy to the square of opposition:

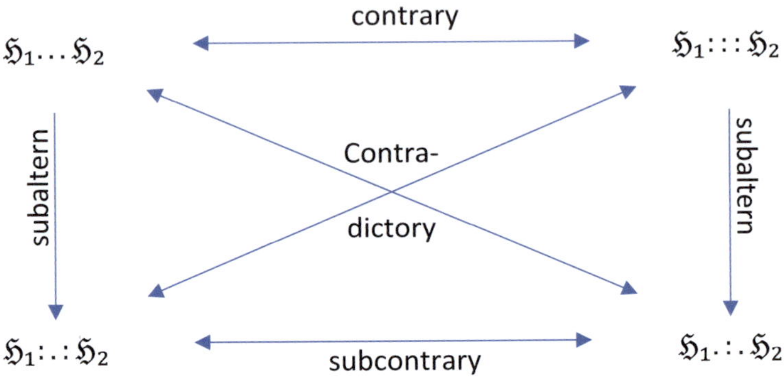

The most interesting relation within the square of world histories is probably the subcontrary: two world histories can be partially overlapping and partially quasi-parallel. Stating only the former leaves room for the possibility that the two world histories in question are entirely overlapping. Adding the latter, i.e., world histories being partially quasi-parallel, eliminates this possibility. Likewise, stating only the latter leaves room for the possibility that the world histories are entirely quasi-parallel. But adding the former, i.e., world histories partially overlapping, again, eliminates this possibility. In other words, combining the partially overlapping case with the partially quasi-parallel case allows us to specify a case where two world histories are overlapping in some regions of configuration space, but parallel in others. With this, we have developed the necessary preliminaries to address the topic of splitting and merging world histories.

4.9.3 World Histories: Splitting and Merging

Let's take a look at the simplest variant of the case just mentioned, with world histories overlapping in one region of configuration space and being parallel in another. Recall Fig. 4.4 above. Figure 4.5 is the same, with a slight change in indices. To make things simpler (and easier to draw), we imagine that in the region where the world histories are properly overlapping, they do so perfectly, such that, in this region, they are identical even on the fundamental level. While this is not generally so, the condition for overlapping world histories, whether entirely or partially, covers this case.

Now, by adding some more worlds, we can turn this into a picture that contains two world histories that are partially overlapping as well as partially quasi-parallel, i.e., that fall under the conjunction of the mutually subcontrary cases, $\mathfrak{H}_1 :.: \mathfrak{H}_2 \wedge \mathfrak{H}_1 .:. \mathfrak{H}_2$ (Fig. 4.6).

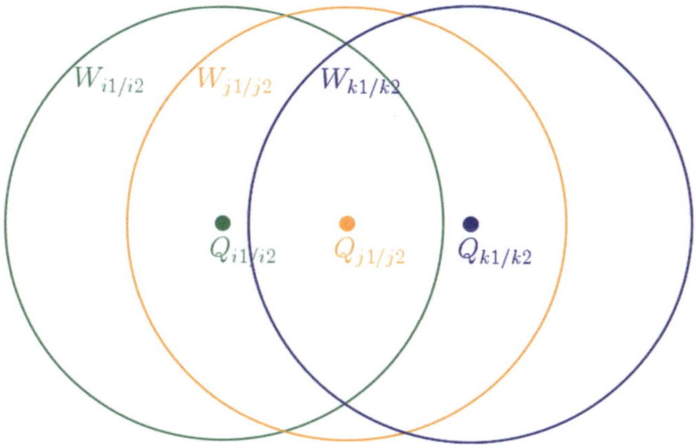

Fig. 4.5 Two (so-far) identical world histories. (Note the indices!)

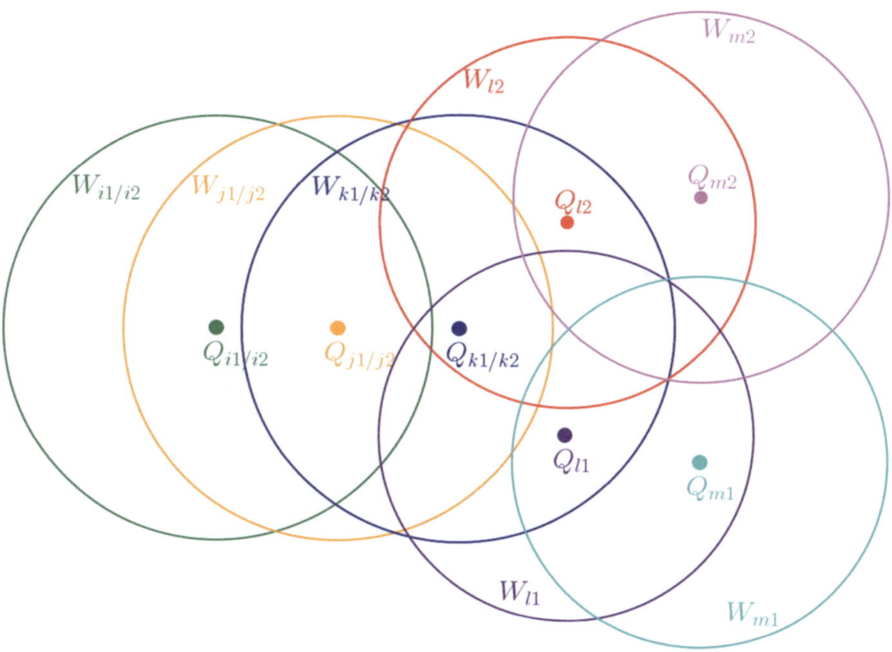

Fig. 4.6 Two world histories, partially overlapping and partially quasi-parallel

Here, Q_{l1} and Q_{l2} both lie within the world of $Q_{k1/k2}$ (and vice versa). As such, they are continuations of the original world history $\mathfrak{H}_1$ (and of $\mathfrak{H}_2$, which, from $Q_{i1/i2}$ to $Q_{k1/k2}$, is identical with $\mathfrak{H}_1$). But they are not similar enough to lie within each other's worlds. Note that, since worlds $\mathcal{W}_{l1}$ and $\mathcal{W}_{l2}$ do not properly overlap, i.e., neither Q_{l1} nor Q_{l2} lies within the world of the other, world histories $\mathfrak{H}_1$ and $\mathfrak{H}_2$ satisfy the condition for partially quasi-parallel world histories. At the same time, they satisfy the condition for partially overlapping world histories, since worlds $\mathcal{W}_{i1/i2}$ through $\mathcal{W}_{k1/k2}$ overlap properly. It is because these two conditions are satisfied that splitting of world histories takes place.

At first sight, there seems to be something peculiar about this picture: read from left to right, it depicts the case of two overlapping world histories (entirely, when looking only at $\mathcal{W}_{i1/i2}$ through $\mathcal{W}_{k1/k2}$), such that they are FAPP one world history,[37] splitting into two quasi-parallel world histories. Read from right to left, it depicts the case of two quasi-parallel world histories, merging into two overlapping world histories. Which way of reading this picture is the right one? Are these world histories splitting or merging?

The issue is immediately relieved once we remind ourselves that TBM imposes a (temporal) order on its actualizations. To be clear: TBM prima facie does not impose a *specific* order on its actualizations. Configurations are actualized at random, i.e., are stochastically independent. Yet, there is only one configuration actualized at a given time, and the next one after that, and so on. Thus, the totality of actualized configurations is ordered, although this order is not predetermined. This ordering carries over to subsets of the totality. Now, focus on configurations that contain MOs and thus instantiate worlds. Since all subsets of the totality are ordered, subsets containing MOs are ordered as well. And since configurations containing MOs instantiate worlds, worlds are ordered, too. If these worlds satisfy the conjunction of overlap conditions as specified above, they form a world history. Therefore, the worlds making up a world history are ordered. We thus know in which direction to read the picture. Assume configurations are actualized in the following order (with possibly other configurations actualized in between): $Q_{i1/i2}$, $Q_{j1/j2}$, $Q_{k1/k2}$, Q_{l1}, Q_{l2}, Q_{m1}, Q_{m2}. From this order, we know that we have to read the picture from left to right. It represents splitting, not merging, of world histories. Note two things: (1) The order just specified is not the only one to instantiate splitting world histories. Any permutation of the last four elements does, if it satisfies $Q_{l1} <_{before} Q_{m1}$ and $Q_{l2} <_{before} Q_{m2}$. That is to say: Within a world history, actualizations of configurations are ordered, but it is irrelevant whether in between these actualizations, configurations from a different world history are actualized or not. Above we have called this external interruption. (2) Ignoring order, there are other possible ways of reading the picture. E.g., starting from $\mathcal{W}_{m1}$ via $\mathcal{W}_{l1}$ to $\mathcal{W}_{k1/k2}$, then splitting into

[37] World histories, in the regions where they entirely overlap, are FAPP one world history, even when they do not overlap as perfectly as depicted in Figs. 4.5 and 4.6. The reason being that entire overlap requires that for every central configuration of one world history, there is another one from the other world history that lies within the former's world. Hence, whenever a central configuration of one of their worlds is actualized, the thus instantiated world overlaps properly with a world of the other world history, such that this instantiation does not amount to macroscopic change for the MO.

two world histories, $\mathcal{W}_{j1/j2}$ to $\mathcal{W}_{i1/i2}$ and $\mathcal{W}_{l2}$ to $\mathcal{W}_{m2}$. This and similar possible readings are rendered impossible by the order specified before (and permissible permutations).

A remark about the ordering of actualizations is called for. One might want to know: what imposes this order? I have said that there is one configuration actualized *at a time*. Does this imply that some kind of external, universal, even absolute time is assumed? That is to say, is the order of configurations given by some kind of unambiguous "time stamp", attached to each configuration as it is actualized, that corresponds to the universal time parameter as it increases? For this presentation of TBM, the answer must be a clear "yes": indeed, we assume absolute, universal time. The time parameter featuring in the time-dependent SE governing the time evolution of the universal WF suggests itself for such an interpretation. Yet, that this assumption is controversial should be clear, too. It is for a reason that Barbour adopts a timeless description of the universe, employing the Wheeler-DeWitt equation. Whether TBM can be altered to be made compatible with such a fundamentally timeless picture of the universe is a question that lies beyond the scope of this book. For now, suffice it to say that the assumption of a fundamentally timeless universe has implications on the construal of world histories. With all configurations actualized eternally, we can no longer refer to an external time parameter for their ordering. Hence, the solution applied to the question of splitting and merging world histories above is blocked in such a universe, unless time somehow "re-emerges" on the level of world histories. But for the time being, let's proceed with our discussion of world histories in TBM, assuming universal time, by applying our framework to the double-slit example familiar from above. This shall prove helpful in making more graspable how world histories emerge from TBM's discontinuous micro dynamics.

4.9.4 An Application of Splitting World Histories: The Double-Slit Experiment

Let's look back at Fig. 4.6, showing the splitting of world histories. We are now going to use it to get a clearer idea of what happens in the double-slit experiment, according to TBM. We start at configuration $Q_{i1/i2}$.[38] When this configuration is actualized, it contains all there is in the universe, at that moment in time, from the large scale filaments of galaxies, spanning over hundreds of millions of light years, individual galaxies, to our solar system and earth, including the laboratory and apparatus set up for our experiment, and the particles the behaviour of which we want to investigate. In particular, it also contains an experimenter, who we choose

[38] That is not to say that the world history really started there. We pick $Q_{i1/i2}$ as a starting point for our considerations because it is convenient.

to be the MO with respect to which the situation is described and worlds are construed. At $Q_{i1/i2}$, ever so slightly, the experimenter still feels the start button for the apparatus at their fingertip. At $Q_{k1/k2}$, they don't feel that button anymore. And the actualizations in between, $Q_{j1/j2}$ as an example taken from the infinity of others, account for the transition between still feeling the button, in $Q_{i1/i2}$, and not feeling it anymore, in $Q_{k1/k2}$. Let's say that, at the point in time when $Q_{k1/k2}$ is actualized, the particle resides within the slit. To be precise: at that point in time, Q_{k1}, with the particle in the left slit, has some disposition of being actualized, and Q_{k2}, with the particle in the right slit, has some disposition of being actualized. (Remember that these configurations are merely drawn in the figure as the same configuration, while actually being different on the micro level.) With TBM's arbitrarily high actualization rate, configurations sufficiently similar to Q_{k1} and Q_{k2} and thus lying in the worlds $\mathcal{W}_{k1}$ and $\mathcal{W}_{k2}$ are instantiated arbitrarily often: the particle jumps back and forth between the two slits (and other places, too). At least, that is what TBM's micro dynamics tells us. For the experimenter, it makes no difference where the particle is at that moment. It could be somewhere near Andromeda. And, assuming a finite probability for the actualization of that configuration, it will be. As long as these configurations don't amount to a noticeable macroscopic difference for the experimenter, their worlds overlap properly. This apparent curiosity aside, at some point, the experimenter will notice a macroscopic difference: Q_{m1} and Q_{m2}, both being continuations of the original world history, but sufficiently different for the experimenter to be macroscopically distinguishable, instantiate worlds where the screen is lighting up at different points. Figure 4.6 conveys the impression that these points should be very close together on the screen, given that the distances between the configurations represented must be miniscule in configuration space. But don't let the limitations of this kind of representation (and of my drawing skills) fool you. You may think of Q_{m1} and Q_{m2} as being as close or far apart as you like, as long as they both instantiate different macroscopic states of the screen, lighting up here or there, and likewise, different macroscopic states of the experimenter. Only with such macroscopic differences splitting of world histories occurs.

Now, think this on into the future. Q_{m1}'s and Q_{m2}'s world histories are continued by the actualization of further configurations. With the experimenter being in Q_{m1}'s world history, say, the experiment goes on, and another point on the screen will light up, leading to branching of world histories again. And the same, of course, for the world history of Q_{m2}. Where the screen lights up in the next event, resulting in the next split of world histories, is just as random as before, "guided" by the universal actualization disposition. But now, the respective world histories already contain records of the first event.[39] And when these world histories split again, the resulting

[39] Already Bell (1987a, 136, his emphasis) notes that memories and records, telling of different times while being contained in one configuration, can be correlated: "We have only our 'memories' and 'records'. But these memories and records are in fact *present* phenomena. The instanta-

world histories will contain records of the first two events, and so on, until each resulting world history, at the end of the experiment, contains sufficient records for an interference pattern to emerge on the screen.[40]

Two remarks are in order. First, note that the fact that these records are present in future instantiations must not be ascribed to the proper overlap of worlds in the respective histories alone. Records, in principle, can vanish in a similar fashion they appear. It is a contingent matter of fact that some records are stable over longer time intervals than others. (The photographic plate that is the screen might vanish after a few centuries; geological formations remain for millions of years; memories of the dreams I have usually last for less than a minute.) Ultimately, stability of records must be attributed to the universal WF and its evolution. However, *if* they are stable, the continuation of world histories via properly overlapping worlds guarantees that they are *reliable*: once a *stable* record is created in a world history, the continuation of this world history will contain this record, too, for as long as it is stable, because worlds that don't contain it are macroscopically distinct and hence not part of this world history. In other words: within a world history, stable records, like those on a photographic plate, do not suddenly vanish. This, although in general, the continuation of world histories via worlds overlapping properly does not impose any kind of (quasi-classical) dynamics onto the universe. This must stem from the dynamics of the universal WF. Proper overlap of worlds *alone* merely guarantees that transitions between worlds are unnoticeable macroscopically, despite a discontinuous micro dynamics, thus explaining the FAPP-continuous evolution of macroscopic processes. In short: world histories take care of *FAPP-continuity* despite a discontinuous micro dynamics of the configuration, while the dynamics of the WF (including decoherence) takes care of *quasi-classicality*.

Second, the example as presented might leave the impression that, on TBM, the experimenter/observer does play a special role with respect to the continuation and splitting of world histories. Yet, this is not the case. They do play a certain role, but only insofar as they are an instance of MOs in general. In particular, it is irrelevant whether the MO in question bears a mind, has conscious experience, etc., or not. We could as well have left out the experimenter from the description, taking the screen itself for the MO. The results with respect to the splitting of world histories would have been decidedly similar, since the observer and the screen are roughly on the same level of macroscopicity: if world histories split for the screen, they split for the experimenter, and vice versa. Then again, viewed from yet another level, on the

neous configuration of the x s can include clusters which are markings in notebooks, or in computer memories, or in human memories. These memories can be of the initial conditions in experiments, among other things, and of the results of those experiments. The theory should account for the present correlations between these present phenomena. And in this respect we have seen it to agree with ordinary quantum mechanics, in so far as the latter is unambiguous.

[40] Interestingly, these interference patterns, although being different in each world history branch, can be expected to be roughly the same with respect to their overall structure. They thus grant us a glimpse into the goings-on of the underlying, microscopic level, despite the fact that the world histories we inhabit only emerge on the macro level, and irrespective of which particular world history we inhabit.

order of the solar system, say, no splitting of world histories results from the experiment as described: for an MO on that scale, it literally makes no macroscopic difference whether some screen in some lab on some planet lights up here or there. But of course, things could be arranged such that it does. If we were to "blow up" the result of such an experiment by linking it to an array of nuclear bombs, literally blowing up Mars, for example, iff the first spot on the screen appears in a defined region, and not iff it appears elsewhere, that would indeed make a macroscopic difference for the solar system. Accordingly, in such a scenario, the result of the experiment would lead to splitting of world histories even for the MO that is the solar system.

What this example is supposed to make clear in a more graspable way are the following three points—and the appropriate conclusion—that have been made before: (1) World histories split. (2) Importantly, world histories split, not when some microscopic superposition is present in the state, but when the superposition is magnified to be macroscopically relevant. (3) What is macroscopically relevant depends on the MO in question. In conclusion, the example shows that splitting of world histories is an emergent feature. It naturally results from the TBM framework, i.e., the Schrödinger dynamics for the universal WF, the interpretation of the latter as encoding dispositions for actualizations of configurations, the stochastic dynamics of the actualized configuration in accordance with that interpretation, and the construal of worlds in terms of regions of sufficient similarity with respect to a given configuration, and the MO(s) within it. Accordingly, if there are different MOs for which different processes are macroscopically relevant—like the screen and the solar system in the example—world histories, upon some event, might split or not, because the same event that is macroscopically relevant for one MO might be not for another.

Our treatment of splitting world histories might be suggestive of their countability. I have previously stated that world histories, like Everettian world branches, are only FAPP-countable. I yet owe an explanation for why that is so.

4.9.5 Counting World Histories

World histories are countable, but only FAPP, i.e., when given a specific context, provided by the selection of an MO. Alas, there is no unique answer to the question which MO to select, so there is no unique count of world histories. Consider the following example: for a given MO, two world histories, $\mathfrak{H}_1$ and $\mathfrak{H}_2$, are entirely overlapping, and a third world history $\mathfrak{H}_3$ is entirely overlapping with $\mathfrak{H}_2$. Even if $\mathfrak{H}_1$ and $\mathfrak{H}_3$ don't mutually satisfy the condition for entire overlap, and even when they are entirely quasi parallel, they are connected via $\mathfrak{H}_2$, similar to the quasi-parallel worlds at the "ends" of an individual world history: $\mathfrak{H}_2$, in this scenario, entirely overlaps $\mathfrak{H}_1$ as well as $\mathfrak{H}_3$. Now, is it fair to assume the existence of thusly connected world histories? In this example: is it fair to assume the existence of $\mathfrak{H}_2$ in between $\mathfrak{H}_1$ and $\mathfrak{H}_3$? On TBM's infinite actualization rate, it is. It is even fair to assume that there are infinitely many world histories instantiated in between any two quasi-parallel world histories, just as there are infinitely many worlds instantiated in between two quasi-parallel worlds. But, as has been pointed out before, two entirely overlapping world histories can be considered, FAPP, one world history. Does this mean, then, that $\mathfrak{H}_1$, $\mathfrak{H}_2$ and $\mathfrak{H}_3$ *together* must be considered one world history? After all, $\mathfrak{H}_1$ overlaps $\mathfrak{H}_2$ entirely, and so does $\mathfrak{H}_3$. Then, one might think, they all can be considered one world history, FAPP. However, this is only so for the MO with respect to which $\mathfrak{H}_2$ is construed. If, instead, we consider the MO of $\mathfrak{H}_1$, we see that, because $\mathfrak{H}_1$ does not entirely overlap with $\mathfrak{H}_3$, $\mathfrak{H}_1$ and $\mathfrak{H}_3$ can be FAPP-counted as two world histories. This, despite the fact that these world histories are connected via $\mathfrak{H}_2$, which overlaps both of them entirely.

And there is more. Recall the example from above: for the MO that is the solar system, there are fewer world histories than for the MO that is me. Just which MO we should select as a reference point to count world histories is a question that has no unique answer. So, whether world histories are FAPP-counted as one or several, depends on the MO in question, the reason being that overlap is spelled out in terms of SS between actualized configurations, which in turn is spelled out in terms of an intransitive relation between the macro states they instantiate—macroscopic indistinguishability—relative to a given MO. With this dependency on the MO made clear, it becomes obvious why it is impossible to provide a unique count of world histories.

Rather, counting world histories is a matter of context. Ultimately, world histories, like the worlds that constitute them, are an emergent feature—they only exist FAPP, from the perspective of an MO. Fundamentally, all there is are random actualizations of configurations, following an actualization disposition encoded in the universal WF. Despite, or rather: because of this, world histories are suitable to explain the manifest image of FAPP-continuous, macroscopic processes emerging from the scientific image of a stochastic, discontinuous micro dynamics.

4.10 TBM Is Compatible with Double Aspect PPP

As argued above, in order to be compatible with a more than minimal variant of PPP, in particular in order to be compatible with double aspect PPP, TBM must allow for macroscopic processes from which what we would call a mind capable of experiences can emerge. Individual configurations, even if they are time capsules, didn't seem to be enough, for the processes required as the physical aspect of mental phenomena, i.e., brain processes, cannot arise from within individual, static configurations. Employing FAPP-continuous world histories, we are able to account for experiences, including experiences of continuous macroscopic change, despite a fundamentally random, discontinuous micro dynamics. Our brains are MOs undergoing macroscopic change within world histories. For the latter, we require pairs of consecutive configurations to be sufficiently similar as to not affect the MO macroscopically by transitioning from one to the other. At the same time, because of similarity being a nontransitive relation, a series of actualizations of pairwise sufficiently similar configurations can have configurations at its far ends that are not sufficiently similar to each other, and thus amount to macroscopic change of the MO. If the MO is a brain, this allows for brain processes to happen, from which an experiencing mind can emerge, as the psychological aspect of the physical process. It is probably worth noting again that we don't need a mind already in place for world histories to form. In particular, it is not necessary that there is a mind (or a higher being, even) comparing configurations in order to see whether they are sufficiently similar or not. Sufficient similarity merely requires that two actualisations do not amount to macroscopic change of an MO. Whether they do or not, as argued, depends only on the MO in question, not on a mind observing it, or the MO itself having a mind. It can undergo change, even in an otherwise completely empty universe, deprived of any mental aspect whatsoever. For example, the screen or photographic plate in the double-slit experiment is an MO that undergoes macroscopic change during the experiment, without itself having a mind, and irrespective of whether an experimenter—a special, experiencing type of MO—is watching or not. While undergoing change, according to the best theories we have, seems to be a necessary requirement for a mind to emerge, neither having a mind nor being observed by a mind is a necessary requirement for undergoing change. We are thus not in need of a mind for world histories to form—TBM indeed seems compatible with double aspect PPP.

References

Allori, V. (2015). Primitive ontology in a nutshell. *International Journal of Quantum Foundations, 1*(3), 107–122.

Barbour, J. (1994). The emergence of time and its arrow from timelessness. In J. J. Halliwell, J. Pérez-Mercader, & W. H. Zurek (Eds.), *Physical origins of time asymmetry* (pp. 405–414). Cambridge University Press.

Barbour, J. (1999). *The end of time*. Oxford University Press.

Barbour, J. (2009). *The nature of time*. arXiv:0903.3489v1 [gr-qc]

Barbour, J. (2020). *The Janus point*. Bodley Head.

Bell, J. S. (1987a). Quantum mechanics for cosmologists. In J. S. Bell (Ed.), 1987b *Speakable and unspeakable in quantum mechanics*. Cambridge University Press.

Bell, J. S. (1987b). *Speakable and unspeakable in quantum mechanics*. Cambridge University Press.

Black, M. (1952). The identity of indiscernibles. *Mind, 61*(242), 153–164.

Bohr, N. (1949). Discussions with Einstein on epistemological problems in atomic physics. In P. A. Schilpp (Ed.), *Albert Einstein: Philosopher-Scientist*. Tudor.

Brown, H. R. (2010). Reply to Valentini, 'De Broglie–Bohm pilot-wave theory: many worlds in Denial?' In S. Saunders, J. Barrett, A. Kent, & D. Wallace (Eds.). *Many worlds?* (pp. 510–517). Oxford University Press

Brown, H. R., Sjöqvist, E., & Bacciagaluppi, G. (1999). Remarks on identical particles in de Broglie-Bohm theory. *Physics Letters A, 251*(4), 229–235. https://doi.org/10.1016/s0375-9601(98)00907-4

Brown, H. R., & Wallace, D. (2005). Solving the measurement problem: De Broglie-Bohm Loses Out to Everett. *Foundations of Physics, 35*(4), 517–540. https://doi.org/10.1007/s10701-004-2009-3

Chisholm, R. M. (1967). Identity through possible worlds: Some questions. *Noûs, 1*(1), 1.

Cowling, S. (2016). Haecceitism. In E. N. Zalta (Ed.). *The Stanford Encyclopedia of Philosophy* (Fall 2016 Edition). https://plato.stanford.edu/archives/fall2016/entries/haecceitism/

Dürr, D., & Teufel, S. (2009). *Bohmian mechanics*. Springer Science & Business Media.

Dürr, P., & Ehmann, A. (2021). The physics and metaphysics of Tychistic Bohmian mechanics. *Studies in History and Philosophy of Science Part A, 90*, 168–183. https://doi.org/10.1016/j.shpsa.2021.09.014

Forbes, G. (1984). Two solutions to Chisholm's paradox. *Philosophical Studies, 46*(2), 171–187.

French, S., & Krause, D. (2006). *Identity in physics: A historical, philosophical, and formal analysis*. Oxford University Press. https://doi.org/10.1093/0199278245.001.0001

Heidelberger, M. (2003). The mind-body problem in the origin of logical empiricism: Herbert Feigl and psychophysical parallelism. In P. Parrini, W. C. Salmon, & M. H. Salmon (Eds.), *Logical empiricism: Historical and contemporary perspectives* (pp. 233–262). University of Pittsburgh Press.

Jaeger, G. (2014). What in the (quantum) world is macroscopic? *American Journal of Physics, 82*(9), 896–905. https://doi.org/10.1119/1.4878358

Lewis, D. (1986). *On the Plurality of Worlds*. Blackwell.

Kim, J. (1993). *Supervenience and mind*. Cambridge University Press.

Maudlin, T. (1995). Three measurement problems. *Topoi, 14*(1), 7–15. https://doi.org/10.1007/BF00763473

Putnam, H. (1981). *Reason, truth and history*. Cambridge University Press.

Reif, F. (1965). *Fundamentals of statistical and thermal physics*. McGraw-Hill Book Company.

Sacks, O. (2017). *The river of consciousness*. Picador.

Saunders, S. (2003). Physics and Leibniz's principles. In K. Brading & E. Castellani (Eds.), *Symmetries in physics*. Cambridge University Press.

Schlosshauer, M., & Camilleri, K. (2011). What classicality? Decoherence and Bohr's classical concepts. *Advances in quantum theory: Proceedings of the international conference on advances in quantum theory, AIP, 1327*, 26–35. https://doi.org/10.1063/1.3567426

Tao, T. (2011). *An introduction to measure theory*. American Mathematical Society.

Valentini, A. (2010). De Broglie–Bohm Pilot-Wave theory: Many worlds in Denial? In S. Saunders, J. Barrett, A. Kent, & D. Wallace, (Eds.), *Many worlds?* (pp. 476–509) Oxford University Press.

van Fraassen, B. C. (1980). *The scientific image*. Oxford University Press.

Wallace, D. (2012). *The emergent multiverse*. Oxford University Press.

Wittgenstein, L. (1953). *Philosophische Untersuchungen*. Blackwell.

Wu, Y.-S. (1984). General theory for quantum statistics in two dimensions. *Physical Review Letters, 52*(24), 2103–2106. https://doi.org/10.1103/PhysRevLett.52.2103

Zurek, W. H. (2002). Decoherence and the transition from quantum to classical – revisited. *Los Alamos Science*, (27).

Chapter 5
Conclusion and Outlook

Zeno's arrow flies. Not in a single, isolated configuration, or world. But in a configuration or world that is part of a FAPP-continuous world history. Like a tangent through a point on a circle is not a property of the point alone, but of the point as embedded into a continuum of points on the circle, the arrow's flight, including its instantaneous velocity, is not a property of the configuration alone, but of it as being embedded in a FAPP-continuous world history. Without a circle, there is no tangent. Without a world history, there is no movement of the arrow.

In this book, I have developed and presented an account of the emergence of FAPP-continuous, quasi-classical world histories from the stochastic, discontinuous micro dynamics of TBM.

For an introduction to TBM, the transition from BM to TBM was described, and their empirical equivalence established. I presented the conditions under which TBM can be classified as a primitive ontology theory, regarded as crucial for the acceptance of a theory as Bohmian by many proponents of BM. While rendering TBM compatible with the PO approach, it must not fall prey to the problem of communication, arising from the two-space reading of BM. After spelling out the desiderata of the PO approach regarding the fundamentality of the primitives of a theory, I argued that to achieve compatibility with the PO approach while avoiding the problem of communication, TBM's wave function must be interpreted nomologically. That, however, is not to say that I commit myself to this interpretation. I merely argued that TBM can be classified as a PO theory, and thus as a Bohmian theory, if so demanded: given a suitable reading of the wave function, TBM satisfies the desiderata of the PO approach. Thus, Bohmians should abstain from rejecting TBM for not possibly being Bohmian.

I have discussed the issue of temporal solipsism, identified by Bell as a serious problem for his "Everett (?) theory"—the historic predecessor of TBM. I argued that Barbour's time-capsule approach, adopted for TBM, does not provide a satisfactory answer to Bell's worry. In particular, his adopting a more than minimal

A. Ehmann, *Worlds in a Stochastic Universe: On the Emergence of World Histories in Tychistic Bohmian Mechanics*, Synthese Library 537,
https://doi.org/10.1007/978-3-032-14632-8_5

psychophysical parallelism between brain processes and experience of macroscopic change, while being a reasonable assumption in light of our current best neuroscience, is problematic for theories relying solely on time capsules, understood as highly structured, individual, static configurations containing consistent records. As a solution, I introduced worlds, and world histories, as key concepts in providing a link between TBM's discontinuous micro dynamics and FAPP-continuous macroscopic phenomena. Worlds, other than time capsules, are coarse-grained regions in configuration space, defined as sets of possible configurations satisfying a nontransitive relation of sufficient similarity with respect to a given configuration containing macroscopic objects. Hence, worlds may overlap. They thus allow for FAPP-continuous, quasi-classical world histories to form.

To first introduce the notion of overlapping worlds, I discussed their applicability to the case of Boltzmannian statistical mechanics (BSM). There, for finite available phase space volumes, partitioned into non-overlapping macro regions of finite, non-zero volume, the discontinuity problem of BSM arises: BSM invariably models change of macro values in such scenarios as jump discontinuities, at odds both with the expectation that macro values should be able to change continuously—stemming from the desideratum that BSM be able to model thermodynamically described target systems—and with the continuous evolution of the micro state. I have argued that overlapping worlds are a reasonable option for replacing the disjoint macro regions of phase space. This move solves the discontinuity problem.

Having thus introduced and developed the notion of overlapping worlds, and demonstrated its applicability to the case of BSM, I adapted the concept for the case of TBM, where a similar problem arises: FAPP-continuous, quasi-classical, consistent histories (as we seem to live in one) must emerge from TBM's discontinuous, fundamentally random micro dynamics in configuration space, thereby avoiding Bell's charge of temporal solipsism. A special requirement for overlapping worlds had to be fulfilled in order for world histories to form: worlds must overlap properly. I argued that TBM's indefinite actualization rate guarantees that they do. Along the way, issues pertaining to the microscopic-macroscopic-distinction, reduction of configuration space, particle identity, impenetrability and haecceity in light of the desideratum of particle number conservation in non-relativistic quantum theories, etc., have been examined. I provided a detailed explanation of how overlapping worlds in TBM form world histories, thereby linking up macroscopically distinct worlds. Thus, the problem of temporal solipsism was resolved.

My hope was that by spelling out in some detail how world histories are formed, TBM might be rendered more acceptable to those having difficulties imagining how the manifest image of the universe we seem to inhabit can possibly emerge from the peculiar scientific image of the universe TBM describes. While working on TBM for some time now, and discussing it with friends and colleagues, I became increasingly convinced that I should at least attempt to make this emergence intelligible. I hope I have succeeded, if only to some extent. Certainly, there are more questions to ask, and to answer, and to answer more extensively and in greater detail, than I could in this book. TBM is a new theory, despite descending from a decades-old ancestor: Bell's Everett (?) theory. Possibly it was because its original inventor—or discoverer, if you prefer—had dismissed his own theory immediately, and because he was such an eminent figure, that it never really took off. Alas, Bell's criticism was based on a misguided identification of worlds with individual configurations. Had he not made this mistake, TBM by now could have been "out there", as a viable alternative, for decades, possibly developed way further than the juvenile stage it finds itself in as of today. However, Bell did not recognize that worlds can—and should—be construed as coarse-grained, and so, for the time being, TBM is still a work in progress, with plenty of topics to be taken up in possible follow-up projects. For an example, there are interesting questions to engage with concerning the personal identity of inhabitants of (splitting) world histories: can persons inhabiting remote parts of world histories reasonably be identified with one another? What about persons in parallel word histories? If the latter are a result from splitting, are there now multiple persons where there was one, before the split? Or have there always been several, already before splitting? What are the differences, if any, between simultaneously realized, Everettian world branches, and TBM's world histories, with respect to personal identity? I think such and related questions might be of interest even to a broader audience, not particularly concerned with TBM in itself. Likewise, issues pertaining to the relationship between world-history internal time and external, universal time may be worth investigating. After all, world histories may be externally interrupted without being internally interrupted, as long as their worlds overlap properly. The fact that, on TBM, there is no necessary link between fundamental, universal time and world-history internal time (except for ordering actualizations) might inform more general debates in the philosophy of time broadly construed. I would like to pick up these and similar topics in future projects, and can only invite the interested reader to join me.

MIX
Papier aus verantwortungsvollen Quellen
Paper from responsible sources
FSC® C105338
FSC
www.fsc.org

If you have any concerns about our products,
you can contact us on
ProductSafety@springernature.com

In case Publisher is established outside the EU,
the EU authorized representative is:
Springer Nature Customer Service Center GmbH
Europaplatz 3, 69115 Heidelberg, Germany

Printed by Libri Plureos GmbH
in Hamburg, Germany